AMERICA'S FIGHT
FOR ENERGY INDEPENDENCE

POWER
STRUGGLES

MICHAEL MAUCELI

A RICH DAD LIBRARY SERIES BOOK
with Foreword by Robert Kiyosaki

POWER STRUGGLES

AMERICA'S FIGHT
FOR ENERGY INDEPENDENCE

WHY OIL AND GAS WILL REMAIN OUR #1 ENERGY SOURCE
— AND HOW YOU CAN PLAN AND PROFIT

MICHAEL MAUCELI

PLATA
PUBLISHING

Published by Plata Publishing, LLC platapublisherservices@richdad.com
Readers can reach Mike Mauceli via: REIEnergy.com

Printed in Canada
First Edition: July 2025

Print ISBN: 978-1-61268-145-0
eBook ISBN: 978-1-61268-147-4

072025

CONTENTS

PREFACE

What a difference a day makes... specifically, in this case, Election Day 2024 in the United States.

The country is poised, it seems, for a massive shift related to energy: From stopping work on the third leg of the XL Keystone Pipeline—as the first act, via Executive Order, of the 2021 incoming President— to today's 2025 new-administration theme of "Drill, baby, drill!"

These decisions, policy decision, could not be more extreme... and more indicative of the forces, and types of forces, that will shape the future related to energy. And U.S. energy independence.

The position of the United States related to energy—sources and resources and agendas—takes me the first of the 'p words' I'll reference in this Preface: Policy.

Although our overarching goals as a nation should be unwavering and laser focused, polities and party agendas continue to play a role in positions, progress... and policy. I get it. There are "many roads to heaven," as they say, and many paths to achieving our goal of energy independence. More 'p words,' it seems… politics, parties, paths, and progress.

Which takes us to power. Yet another 'p word'—and what lies at the heart, many would say, of the policies that drive economic, social, and environmental agendas. It's a power struggle, without a doubt. And it's not unreasonable to wonder, and ask, who the winners will be. The American people, one would hope. As opposed to those who hunger for power for power's sake alone… and who see energy as the true power (on many levels) that it truly is. We The People, it often seems, are caught in the middle… trying to come to grips with rising energy costs, depleted oil reserves, jobs in the energy sector, and what all of this means to our future and security and peace of mind. It's stressful to be at the mercy of the 'powers that be' and subject to their agendas and policies—and the ramifications of both.

I'm a guy who believes in controlling what you can control. That can be a tall order in a world where, more often than not, it seems like more and more of the things that impact our lives and our livelihoods— and our futures—are out of our hands. Out of our control.

Growing up in the Delta of Mississippi, where opportunities were scarce, I was fortunate to have parents who were entrepreneurs. Entrepreneurs who knew that they could build a future for their family, work to capitalize on opportunities, and take control of their choices and their investments.

I wrote *Power Struggles* as a way for people to learn about the energy that fuels our future and how to ride the wave of policy decisions that seek and harness opportunities that may be on the horizon. There isn't one of us who has a crystal ball that will give us a glimpse into what the future will actually hold. We can only train ourselves to be more aware of the world around us as it relates to the future of energy and learn how to see opportunities… opportunities that many may miss.

That ability, I've come to learn—after decades in the oil business—is true power.

FOREWORD
BY ROBERT KIYOSAKI

Believe it or not, my college degree is in oil… not finance.

In 1969, I graduated from the U.S. Merchant Marine Academy with a BS degree in Ocean Transportation… with a minor in oil. I studied the naval architecture of oil tankers and the oil industry.

In 1969, my very first real job was with Standard Oil of California. I was a third mate, on board their oil tankers, delivering oil from California to Hawaii, Tahiti, and Alaska.

In 1969, my first port of call as a ship's officer was Valdez, Alaska, the end of the line for oil flowing from the North slope of Alaska.

In 1989, the Exxon Valdez, a supertanker operated by Exxon Oil, struck a reef near Valdez, Alaska, spilling over 10 million gallons of oil over one of the most beautiful areas in the world… Prince William Sound.

In 1969, our Standard oil tanker avoided the same reef the Exxon supertanker ran aground on… 20 years later.

That was good news, certainly, and even better news for me was that, by 1989, I was no longer a tanker officer. I had served six years as a U.S. Marine Corps pilot, flying helicopter gunships… fighting for—you guessed it—the oil in Vietnam. Occidental Petroleum, a huge U.S. oil company, controlled the massive oil reserves in Vietnam… which the Chinese wanted.

By 1989, I was out of the oil business and no longer a pilot. In 1989, I was an entrepreneur starting businesses, investing in multi-family apartment properties, and oil.

Also in 1989, millions of hippies and 'greenies' were up in arms demanding the end to the transportation of oil and gas by ship… proving how little they knew about oil, much less the importance of energy to any civilization.

The Difference Between Stupidity and Ignorance

Most hippies and greenies are not stupid. I would guess that most care deeply for our environment and our planet's future. I would also guess most of them have at least a high school diploma, many a college degree, and a few advanced degrees from reputable institutions of higher learning.

What I am saying is that most are ignorant, not stupid, when it comes to the roles of oil and energy in our world. Most greenies choose to ignore the subject of energy, which places them in the category of ignorant, not stupid.

And this is the reason why Mike Mauceli's book is important… as a resource for moving from ignorant to informed about oil and energy and how we all must play a role in the stewardship of our environment.

In the interests of transparency and full disclosure, I do invest in oil and gas projects with Mike. I have invested millions with Mike over the past decade, made millions, and—thanks to government tax incentives—those investments are rewarded with tax breaks that mean paying less in taxes.

Energy and Civilization

Why is Mike Mauceli's book important?

I can answer that question with four words: Civilizations run on energy. When the price of energy goes up, inflation goes up… and civilization begins to die.

In 2021, the very first act of President Joe Biden when he took office was to shut down funding for that third 'leg' of the Keystone XL Pipeline project. That project brought oil from Canada into the United States and refineries in Texas.

The moment Biden and Harris cut the Keystone Pipeline project, the price of oil shot up.

Under President Donald Trump, Mike and I were selling oil for about $30 a barrel. When Biden and Harris killed the Keystone XL Pipeline deal, the price of oil shot up almost immediately—to $130 a barrel. We are among those who were pleased to see President Trump reverse Biden's actions related to the Keystone XL Pipeline and his focus on domestic energy production.

While I was getting richer as an investor, we were all paying higher prices for fuel and groceries… and virtually everything. Inflation took off and the poor and middle class, especially, were most impacted and getting poorer by the day. Energy drives our world and when *it* gets more expensive it triggers a domino effect... and *everything* gets more expensive.

I remember sitting at a gas station, filling up my car, watching a "soccer mom" with an SUV filled with kids, shaking her head as the cost to fill up her gas tank kept going up and up. I suspect she was trying to figure out how to afford to feed her kids, get them back and forth to school and soccer practice... and keep a roof over their heads.

I remember Federal Reserve Bank Chairman Jerome Powell, in 2000, proclaiming, "This inflation is transitory." I suspect he was delirious... or lying... or ignorant. And immediately I smelled a rat.

When I was attending Kings Point, one of America's five federal military academies and where I studied naval architecture and oil tanker operations, one of the books my economics teacher had us read was Karl Marx's book, *The Communist Manifesto.* It is a small book, first published in 1848.

These federal academies—U.S. Military Academy at West Point, New York; U.S. Naval Academy at Annapolis, Maryland; U.S. Air Force Academy at Colorado Springs, Colorado; U.S. Coast Guard Academy, at New London, Connecticut; and U.S. Merchant Marine Academy at Kings Point, New York—all have a military affiliation. They are very different than "snowflake" universities where we're seeing Marxist undercurrents and where indoctrination is replacing education.

At Kings Point, our motto was: *"Acta Non Verba."* Deeds Not Words.

We were trained to observe what a person *did...* and not be fooled by their words. Words which can be misinterpreted or qualified or dismissed. We've all heard this adage: Actions speak louder than words. They do. And Kings Point memorialized that mantra: Acta Non Verba.

Observing that Biden's first official *act* as President of the United States, in 2021, was to kill the Keystone XL Pipeline… I knew he was a Marxist.

Not long after the Keystone Pipeline Project EO, Biden signed another Executive Order, this one taking down the wall at our Southern border, a national security project that had become known as 'Trump's Wall.' [Also in the interests of full disclosure, Donald Trump and I have written two books together and I have huge respect for his ability to get things *done*. Once again: acta non verba.]

That Biden-Harris action has allowed tens of millions of illegals— many hard-core criminals and terrorists—to enter America. This invasion and infiltration has already proven fatal to many Americans… and these millions of illegals have stressed our healthcare system, our education system, destroyed some of our most magnificent cities, and are likely giving terrorist organizations new cell footholds in America.

In the 2024 U.S. election, Americans used their votes to voice concerns about both border security and the economy. And energy was cited, time and again, as a means to positively impact the economy. Energy will fuel our future and the future of energy hinges upon informed and visionary strategies for the harvesting and use of *all* types of power.

I mentioned earlier that President Donald Trump and I wrote two books together. We did that because we both stand for freedom, especially financial freedom.

If you desire more personal financial freedom, Mike's book *Power Struggles* is essential to your freedom—and the freedom of the world.

Those in the energy sector, like Mike, would use these words as one definition of freedom: energy independence. It is within our reach.

Thank you.

Robert Kiyosaki

Author of the international bestseller *Rich Dad Poor Dad*

INTRODUCTION

"Courage, determination, and hard work are all very nice, but not so nice as an oil well in the back yard."

— WRITER AND APHORIST MASON COOLEY

There's an old joke that goes something like this:

> An old Southern country preacher from west Texas had a teenaged son. The preacher felt that the boy was old enough that he should start giving some thought to choosing a profession.
>
> However, like many young men, the boy didn't really know what he wanted to do for a living, and he didn't seem too concerned about figuring it out.
>
> So one day, while the boy was away at school, his father decided it was time to try an experiment. He

went into his son's room and placed four objects on the boy's desk: a Bible, a silver dollar, a bottle of whiskey, and an issue of Playboy magazine.

"I'll just hide behind the door," the old preacher said to himself, "and when he comes home from school this afternoon, I'll see which object he picks up. That will tell me what the boy ought to be when he grows up!"

As the preacher waited, he thought, "If he goes for the Bible, he'll be a preacher, just like me! What a blessing that would be!

"If he goes for the silver dollar, he's going to be a businessman, and that would be OK.

"But if he picks up the bottle of whiskey, he'll be a no-good drunkard, and Lord, what a shame that would be!

"Worst of all," the man finally thought with disdain, "if he picks up that magazine, he's gonna be a skirt-chasin' bum!"

The old man waited anxiously, and soon he heard his son's footsteps as he entered the house, whistling, and made his way to his bedroom. The preacher watched as the boy tossed his books on the bed, and as he turned to leave the room, he spotted the objects on the table. With curiosity in his eyes, he walked over to inspect them more closely.

Finally, the boy picked up the Bible and placed it under his arm. He picked up the silver dollar and dropped it into his pocket. He uncorked the bottle and took a big slug of whiskey while he admired that month's Playboy centerfold.

After the boy left the room, the preacher exclaimed in a disgusted voice, "Lord have mercy! He's going into the oil business!"

This old chestnut may be funny, but it still rings with truth. That perception of the unsavory oil man—the greedy land-gobbler who drinks and ogles women—still persists a bit today, as does the perception of the oil business as greedy, destructive, and unethical.

Yet as I look back on my nearly 45 years in the oil and gas industry, I've seen remarkable changes in how we do business, how we treat the land, how we use technology to explore for and extract oil from the ground, and how people around the world use oil and gas. I've seen this country go from being starved for oil, on the heels of an Arab oil embargo that nearly brought us to a standstill, to being almost entirely energy-independent, to seeing the industry nearly destroyed in the name of climate change.

And despite the persistent belief that the oil industry is full of corruption, the fact is that it's one of the most heavily regulated sectors in our economy. According to the RegData Industry Regulation Index, on the list of the 10 most regulated industries in the United States as of February 2022, oil and gas extraction ranks at No. 8, and petroleum and coal products manufacturing tops the list at No. 1. The sector faces tough scrutiny with regard to the measures it takes to protect the environment, including air and water quality, and it is tightly regulated for its worker safety practices, among many other issues.

Not only that, but the oil and gas industry has been a key driver of our economy.

As the United States rebounded from The Great Recession, the greatest amount of job creation was in the energy sector, particularly in oil and gas. Between 2007 and 2012, when most industries were shedding workers daily, employment in the oil and gas industry—in drilling, extraction, and support jobs—rose by 40 percent. This contributed significantly to the nation's economic recovery.

Despite a slowdown in the last decade, including during the COVID-19 pandemic, jobs in the industry are still abundant, and salaries have grown with them, according to an August 2023 *Wall Street Journal* article by Mari Novik. There are currently about 937,000 active wells in the United States, reported the Congressional Research Service in January 2023. Oil-and-gas companies posted record profits in 2022. And *Oil & Gas Journal* announced, "The total number of jobs in 2027 is expected to hit 1.09 million," which is an increase over pre-COVID numbers and growth is expected to hit 40% compared to 2020.

> Between 2007 and 2012, when most industries were shedding workers daily, employment in the oil and gas industry—in drilling, extraction, and support jobs—rose by 40 percent. This contributed significantly to the nation's economic recovery.

Advances in technology have not only brought up the production of oil and natural gas, but this new technology is cleaner, more precise, and more effective at pulling crude out of the ground quickly. And despite the highly touted advances in alternative energy development—something that we can all agree is important and inevitable—these innovations have been surpassed by those in the oil and gas business, which now incorporates such advancements as the Internet of Things (IoT) big data analytics, cloud technology, predictive maintenance, and artificial intelligence.

Thanks to these innovations, we can now tap into and recover the tremendous amount of reserves and shale oil in this country—reserves that helped us become energy independent in 2019, finally producing more oil than we were using in this country, according to *Forbes*.

As if all this isn't enough good news, there's more. And it's the reason I've written this book: Investing in oil and gas right now can provide incredible cash-flow opportunities, making it one of the soundest and most profitable investments you can make.

As Robert Kiyosaki likes to point out, investing in real estate or oil and gas (which operates like real estate in many ways) is much safer than taking a ride on the unstable stock market—it's a hedge against inflation, and it's one of the easiest ways to generate cash flow. After all, people will always burn oil and gas and need a roof over their heads. A report by the BDO Alliance, a network of accounting and consulting professionals, projects that the U.S. oil-and-gas industry will experience increased revenue from expanding markets, natural gas will comprise 50% of power generated in the nation, and the United States will become the world's largest exporter of liquefied natural gas.

That's why there's so much institutional money being poured into oil and gas right now. And thanks to recent changes in the tax laws, investing in oil and gas just got even more attractive, because benefits earned through real estate and oil and gas are virtually untaxed. In fact, the benefits of oil and gas actually *exceed* those of real estate. The federal government *wants* you to invest in oil and gas, and it has designed laws specifically to encourage it.

In this book, I'll share you with the exciting changes I've seen since my first day working in the oil-and-gas industry 45 years ago. I'll tell you about the important changes that have been happening in oil and gas in the last few years, and what that means to this country. I'll deal with some of the myths and misunderstandings about this industry that are prevalent here in America. And I'll show you how investing in oil and gas works, and why it's one of the smartest ways you can protect and grow your money.

CHAPTER 1

THE BIG MOVE

"Where oil is first found is in the minds of men."

— AMERICAN PETROLEUM GEOLOGIST WALLACE PRATT

I was just 18 years old on the steamy June day in 1976 when I began my career in the oil-and-gas industry.

I had graduated high school the year before, and up to that point I hadn't accomplished much besides fishing along the shore of Lake Ferguson. As my mother pointed out with increasing frequency, it was time for me to grow up and get a job. But in my hometown of Greenville, Mississippi, there weren't all that many opportunities for an 18-year-old, and I couldn't laze around all summer just waiting for my college classes to start. I needed to start behaving like a responsible adult!

In those days, I often fished the days away with my first mentor and fishing buddy, my Uncle Tony. Tony had worked for the Tennessee

Gas Transmission Company, or Tenneco, for 30 years, and managed to pull a few strings to get me an entry-level summer job.

Tenneco operated three 57-inch natural gas transmission pipelines that delivered natural gas from the Gulf Coast of southern Louisiana, through Greenville, all the way to Chicago, Illinois, where it was sold or distributed all along the East Coast. Along the pipeline route, a series of compressor stations, or pumping stations, were located every 40 to 100 miles. At these stations, the natural gas was compressed, pumping up its pressure and thereby propelling it along the pipeline to its destination.

My first assignment as an employee for Tenneco: Go to the south pasture behind one of these football field-sized compressor stations and examine that section of pipeline for leaks.

As they often say in the corporate world, you have to start at the bottom and work your way up. And it didn't get any more bottom-level than this.

First, I had to cross a pasture lined with bulls tied to a long cattle trailer—it was bull castration day, and I was an unsuspecting witness. As if the horrific sights and sounds of that event weren't enough, I was about to enter my own kind of hell. On this miserably humid midsummer day, wearing a long-sleeved shirt and slathered in Noxzema, I was forced to work six feet underground. This was where the pipeline sat, and where I would spend my day wiping a toxic, black, tar-like substance called cresol, which covered the pipe and burned any skin it came into contact with (hence, the Noxzema), off the pipeline with a huge knife. This enabled another worker to come in with magnetic particle inspection equipment and check for leaks along the pipeline.

Apparently, I have a weak mind and a strong constitution because I managed to spend three summers doing this sweaty, physically taxing work, in between my years at college. Then, during my junior year, Tenneco offered me a full-time position. It was a generous offer, one I might have accepted if my plans had been to remain in Greenville.

As it turned out, however, I was headed to Texas, for two reasons. The first was that during my junior year, I had fallen hard for a girl, skipped out on about half of my semester to be with her, and nearly flunked out of school. In truth, I hadn't really enjoyed school to begin with and was somewhat bored with the whole experience. I was anxious to get started in life and begin a career in business.

I'd hailed from a family of entrepreneurs. My grandfather had owned a liquor store, my father had owned a vending machine business, and my mother had owned a real estate firm. The examples they had set for me illustrated the value of hard work and a solid work ethic. Partying and goofing around with girls while my parents paid for my education simply wasn't in the cards.

And that's what brings me to my second reason for my move to Texas. As you can imagine, my mother was none too pleased about my lack of commitment to school. On the day I arrived home in Greenville in May 1978 to begin my summer break, my mother, disappointed by the report card she'd received in the mail, came out to my car to greet me.

"There comes a day when every little bird must fly," she told me as she stood in the driveway, "and today is your day. Don't bother unpacking your car." She reached out and handed me a $100 bill, then said, "You have a choice. You can either move to Texas to live with your brother in Houston, or you can join the Army."

So off I went to Texas, and it turned out to be the best decision I ever made.

Finding My Calling

When I first arrived, I approached the Electrolux vacuum cleaner company, where my brother worked and my cousin was executive vice president, and secured myself a sales job. Growing up with parents who were both experts at sales, I'd learned a thing or two about how to sell, and I was pretty good at this job.

But I knew even more about the oil-and-gas industry and saw myself with a future in it. After working for Electrolux for about a year, I answered an ad for a job with Golden Eagle Oil & Gas.

At the time, in early 1980, the Bureau of Land Management was offering the Oil and Gas Management Program, one of the most important mineral leasing programs to ever have been offered by the federal government. This program, which essentially was a lottery, enabled individuals to bid on parcels of property that were part of an active oil play in Wyoming's Green River Basin.

Lucky winners would, of course, find themselves in the enviable position of being wooed by oil companies, offered leases complete with lease bonuses (money paid to mineral rights owners in exchange for granting a lease) and rental or royalty payments, and receiving monthly payouts for as long as the leased property produced oil, if indeed the land was found productive in the first place.

Golden Eagle was a small, independent oil-and-gas company that was raising capital from investors in order to make bids on leases from individuals who had secured parcels in the lottery. Its team of geologists and engineers would go through the parcels put up for bid each month, determining which parcels were most promising in terms of producing oil and gas. Then its investment groups, comprised of individual investors willing to invest anywhere from $10,000 to $100,000 each in the partnership, would bid on the parcels together, share upfront royalty payments offered by major companies, and earn interest if the land performed well.

The job I had applied for and landed involved seeking potential investors and putting these partnership groups together to bid on roughly 500 parcels of land. My knowledge of the oil-and-gas industry, combined with my sales experience honed at Electrolux, made me quite good at this job, and it provided me with a foundation that proved quite valuable to me later in life at my own company.

By this time, the oil embargo of the '70s had come to a close, and we were in the midst of an oil boom. Oil-and-gas companies were thriving, and jobs were plentiful. During this time in the early '80s, I took a series of jobs for different oil companies that could make me increasingly better offers, starting with a position raising money for Western Continent Oil Corporation, a more traditional oil company similar to the one where I had gotten my start, which was drilling wells in north central Texas. From there I went on to work for several Western Continent ex-pats who had formed their own small, independent oil company. Shortly afterward, in 1983, Southern Alliance Oil and Gas offered me a great position as vice president, putting me in charge of raising capital.

At this point in time, oil companies couldn't get financing any other way. The price of oil had dropped, and banks and commercial lenders weren't loaning money to oil companies. Private individuals were the only source of capital out there, and I could see that this was going to be the way of the future. Now, 30 years later, I'm more certain than ever that knowing how to raise capital is the most important business skill one can possess, and I'm thankful every day that this is the path I chose to pursue.

I'm even more thankful of this decision because it introduced me to my future business partner and mentor, Vearl Sneed.

When the Student is Ready, the Teacher Appears

A native of North Dakota, Vearl Sneed lived in the Los Angeles area and was an investor with Southern Alliance. He had been very successful in the real estate construction business, having owned and operated numerous multi-family apartment complexes—more than 500 units—in LA, which he had built from the ground up. A widower who loved investing in oil and gas for the enormous tax benefits, Vearl had a sharp mind that always wanted to learn and grow, and he got bored easily if he wasn't keeping busy.

As VP in charge of raising capital, I spent a lot of time with Vearl on the road, visiting new wells being drilled in Texas and showing him his investments. We got to know each other pretty well in those days, and I'm fortunate that he saw something in me, a young man in his twenties, that he liked. From him, I learned a tremendous amount about how to run a business, how to raise money and secure investors, how to leverage investments to maximize profits and tax benefits, and so much more.

In 1987, Vearl approached me with an offer that changed my life. He'd seen me talking to engineers, geologists, and drillers, and he could see that I really knew the oil-and-gas business… *and* raising capital. He proposed that we start an oil-and-gas exploration company together. He'd put up the money, and I'd run the business.

"I'm only going to offer this once," he told me. "So decide by tomorrow and let me know."

Now, when you're in your twenties and making good money as the VP of an oil-and-gas company, it's tough to consider scrapping all that to start a business—particularly a business in an industry that had hit its peak in 1980 and offered a product whose price was declining rapidly. There was now a glut of oil companies. Was it a good idea to enter an already-crowded marketplace?

But despite the security of my position and my misgivings about what was happening in the industry, I was, at heart, an entrepreneur. This was a once-in-a-lifetime opportunity, a milestone moment in my life when I knew I was headed for great things. I accepted Vearl's offer.

So Vearl and I started REI Exploration Inc. (REI) in September of 1987, with me initially working out of a one-office executive suite. I was not new to working hard and small spaces; when I had begun at Southern Alliance Oil and Gas, another start-up company, I had shared an office with the company's printer.

When I first began raising capital at REI, it was very difficult for me to grasp the concept of asking for money from individual investors. Of course, by then I had gained experience making such pitches to investor groups, but their whole reason for being was to hear such pitches; that had never felt as awkward as these initial one-on-one conversations I was now required to have. In fact, I was afraid to ask the closing question. I guess I was like most new salespeople, afraid of rejection and uncomfortable looking people in the eye and asking them to part with their own money. Oil-and-gas investment, after all, always involves risk, and approaching people one on one about it was intimidating.

Then I realized that investors made investments for a different reason. Don't get me wrong, no one likes losing money. And everyone wants the potential for a great return on their capital. But one of the best things about investing in oil and gas offers is the great tax benefits, which we'll discuss later in this book. Once I learned not to think out of my own pocketbook and ask for more than I thought I was going to receive, I soon became the best salesman in the company.

The first oil well we participated in was located in south Louisiana. We took a 25% interest in one well with several other oil-and-gas companies. The well turned out to be a dry hole—certainly an inauspicious way to start a new company. However, there was a lesson here. I believe the oil business is the most humbling business around, and this taught me to be as confident as possible in the well before involving investors—we had to do a lot of homework. The industry also attracts extremely colorful characters, both good and bad, and we would have to be very cautious about the people we worked with. Everyone has an angle on how to find oil and gas.

Building your own team of people on whom you can absolutely rely is crucial if you are going to succeed. These must include geologists, geophysicists, land professionals, and engineers—all highly educated individuals who can make or break your company.

Also critical are your geochemical, radar, and satellite imaging, which are useful tools in locating oil-and-gas deposits.

The tools and specialized knowledge involved in this industry are actually huge factors in why I love what I do so much. I find the science unendingly fascinating, and, as you'll see later on in this book, the technology makes what used to be impossible, commonplace.

A geologist is a scientist who studies the solid and liquid matter that makes up the earth, as well as the processes and history that have shaped it. Geologists usually engage in studying geology. The work of geologists usually incorporates physics, chemistry, and biology, as well as other sciences.

In the oil-and-gas industry, the work of geologists is crucial, as they provide valuable insights as to what certain areas of earth are comprised of and where oil deposits can be found.

CHAPTER 2

FROM BOOM TO BUST... AND BOOM AGAIN

"Saudi Arabia appears devoid of
all prospects for oil."

— ATTRIBUTED TO A DIRECTOR OF ANGLO PERSIAN
OIL COMPANY IN 1926

These days, it's hard to imagine a time when gas prices were too low. But it's true, and for this reason, it was a pretty risky move to form REI Exploration in September 1987.

To understand what the industry was like in 1987, I should explain a bit of history.

The Organization of Petroleum-Exporting Countries, or OPEC, which had formed in Iraq in 1960, was comprised of Arab nations that were enraged over the fact that land had been taken from Palestine to form the state of Israel. They began to launch a war against Israel and were deeply angered when western nations in Europe and North America began supporting Israel. In 1973, on Yom Kippur, the holiest day for Jews, the Arabs attacked.

OPEC member countries collectively supply about 40% of the world's oil supply and control about 80.4% of the world's total proven crude reserves.

Source: OPEC Annual Statistical Bulletin 2022

King Faisal of Saudi Arabia realized that the Arab nations had considerable control over western nations, in the form of oil; after all, this was where the majority of our oil was produced. He instituted an oil embargo to punish the western nations who had provided arms to support Israel—particularly the United States. From 1973 to 1974, OPEC nations stopped all exports to the United States and other western nations. And this move hurt us even more than they had ever expected.

After the embargo, daily shipments of oil from the Middle East dropped from 1.2 million barrels to a miniscule 19,000 barrels.

Source: "OPEC Oil Embargo." Gale Encyclopedia of U.S. Economic History.
Ed. Thomas Carson and Mary Bonk. N.p., n.d. Web. 15 May 2012

Prices skyrocketed and supplies dwindled. In his book, *How We Got Here: The 70s*, David Frum describes how cars lined the streets waiting for the trickles of gasoline that might be allowed at the pumps:

> "The immediate results of the Oil Crisis were dramatic. Prices of gasoline quadrupled, rising from just 25 cents to over a dollar in just a few months. The American Automobile Association recorded that up to twenty percent of the country's gas stations had no fuel one week during the crisis. In some places drivers were forced to wait in line for two to three hours to get gas."

Realizing that they held all the power over the United States, the world's largest consumer of oil, OPEC lifted the embargo in 1974 while continuing to charge outrageous prices for the product, causing those nations' profits to hit the roof. Prices continued soaring, reaching $1 for a gallon of gasoline for the first time ever in the fall of 1979.

As a result, Americans started being more thoughtful about how they consumed oil. The Nixon Administration offered tax incentives to those developing alternative energy sources. Oil was stockpiled for another rainy day, and production here in our country went into overdrive as America started preparing for further embargoes. The government asked for voluntary cutbacks on gasoline consumption and sales, and many gas stations opted to close on Sundays, or to cap the amount of gasoline they would sell to one customer at a time. Americans started implementing conservation measures, such as turning down their heat, seeking out energy-efficient appliances, and trading in gas-guzzler cars for compact, fuel-efficient ones.

It's pretty clear where all this was leading. What do you get when you combine post-embargo overproduction, slowed usage of oil and gas, and inflated prices? A bubble.

It collapsed in March of 1986, when the price of oil fell from $23.29 per barrel in December 1985 to $9.85 in July 1986. By November 1986, the average price for a gallon of gas was 84 cents.

Building from the Ashes

Even though prices had rebounded to above $15 a barrel by 1987, when REI was formed, everyone in the industry was still smarting from the collapse, and we knew things would never be the same. In those days, a lot of us oil men drove around with bumper stickers on our cars that read, "Lord, give me one more oil boom—I promise I won't screw it up this time."

When the price of a barrel of oil had been at record highs, it may have been tough for the American driver, but it had been a good time to be an American oil man. And after the collapse, there were a lot of us out of work and closing up shop. There was no money to be had to put together land deals, and there was no money to drill.

Not only that, but those in the U.S. oil-and-gas industry believed that all the big American oil fields had already been tapped. And contrary to what you might think, the Arab Oil Embargo hadn't lessened our dependence on foreign oil. In fact, at that time, we were importing more oil than we were producing, from countries all over the globe. The major oil companies—Shell Oil and the like—had their major stakes overseas; after all, there wasn't a whole lot of undiscovered territory here in the states… at least, nothing large enough to make it worth their while to drill here.

That left room for small, independent companies like REI to step in and begin exploring properties.

The petroleum industry is generally divided into three major categories: upstream, midstream, and downstream.

For an analogy of the process, think of it in terms of another natural resource: agriculture. The farmer would be your upstream sector. A farmer is responsible for sowing the seeds, analyzing the soil, irrigating crops, and pulling the harvest out of the ground. In the oil-and-gas industry, companies in the upstream sector handle the exploration—identifying reserves in the ground and drilling down to "harvest" the oil.

The midstream sector is where I began my career in the industry. It begins where the upstream leaves off—in its simplest terms, the midstream industry can be described as the part of the process that involves the shipping and storage of the oil and natural gas. The midstream sector is all about taking the crude oil and natural gas retrieved from the upstream sector and getting it to the downstream processing facilities so that it can be turned into the various finished products in consumers' daily lives.

Much like distributors in the food industry work with farmers to purchase produce and resell them to restaurants and grocers, the midstream system in the oil-and-gas industry collects wet natural gas from the wellheads and transports it to a gas-processing plant via pipelines. It can range in size from a very small system, where gas is processed close to the wellhead, to one that contains thousands of miles of small-diameter, low-pressure pipes that collect gas from hundreds of wells. At the gas-processing plant, methane (otherwise known as dry natural gas) is separated from wet natural gas, leaving natural gas liquids (NGL) as a by-product. NGLs are the heavier elements of wet natural gas—ethane, propane, butane, isobutene, and other condensates. The way midstream companies make their money is by fractionating, or breaking apart, these NGLs, transporting them, and marketing them to those who distribute natural gas, otherwise known as the downstream sector.

Companies in the downstream sector are a lot like grocery stores and restaurants that sell food directly to consumers; they sell and distribute natural gas or petroleum products, including liquefied petroleum gas (LPG), gasoline, jet fuel, diesel oil, other fuel oils, asphalt, and petroleum coke, which is a solid by-product that may be burned like coal to produce energy. This is the sector that interacts directly with consumers.

It is the upstream sector where REI operates. This is the exploration and production (often referred to as the "E&P") sector. It involves the search for, recovery, and production of crude oil and natural gas. And it's this sector where the investment opportunity really lies.

In most areas of the world, a country's natural resources—valuable rocks, minerals, oil, or gas found on or within the earth—belong to that country's government. In these places, the only way organizations and individuals can legally extract and sell those commodities is by obtaining permission from the government.

But in the United States, whoever owns the land owns the resources on, above, or within it. In this most basic type of private ownership, called "fee simple estate," the owner controls the surface, the subsurface, and the air above a property.

Vearl and I took advantage of this economic opportunity and set about finding privately owned, potentially oil-rich properties in Texas that we could lease, thereby gaining access to those resources.

The first step was determining where to drill—which areas had the greatest potential as sources of oil. We set our sights on the fossilized ancient coral reefs in north Texas, a plentiful source of oil.

I'd worked with several engineers, geologists, and other specialists that I'd liked in my previous positions and who were now unemployed. At the time, there was a joke about this: "How do you find a geologist? Go to a restaurant and yell, 'Waiter?!'"

I took advantage of this. I started contacting people in the pool of available talent and talked to them about this new opportunity. Many of them had uncovered prospects while they had been employed by the major oil companies—working up drilling site prospects, based on geological data, and examining the quality of the ground in those areas. Then, when they'd lost their jobs, they'd put those prospects in their files for a rainy day. Many had gone into business for themselves, acting as brokers by offering up those prospects to oil companies, and many could make a nice amount of money doing so.

The first thing I did was reach out to a geologist from Abilene named Billy Morris, who had significant experience with, and knowledge about, the land in central and west Texas. He agreed to work with us, and with his prospects in hand, we in turn presented them to potential investors. With a pool of investors committed to a project, we could enter into a lease on that property and begin drilling for oil.

There are numerous ways of financing oil and gas projects today, which I'll get into later in this book. However, in 1987, there were very few options. One of the only methods available to us was to raise capital from wealthy individuals who were looking for a return on their investment and to capitalize on the Intangible Drilling Costs (IDCs) tax deductions offered by the federal government to promote oil-and-gas exploration.

Financially savvy and wealthy people know that the best time to invest is when the market is down—prices are low, and there's nowhere to go but up. That was the case in 1987 when we were offering investments in oil-and-gas development. Property leases were cheap, there was very little competition, the cost of drilling was next to nothing, and many of the smartest minds in the industry were out of work. So as long as we could convince people that oil and gas had a future—and most people understood its value inherently—we could sell it. Investors realized that the industry would recalibrate itself. So even though the industry was at one of its lowest points in history, and many people thought we were crazy, it actually was a really great time to start a domestic oil-and-gas company.

That's not to say it was easy. It was only Vearl and me in the beginning, and though Vearl had put up the money for the venture, he was in his 70s and wasn't interested in working around the clock. I was the one primarily responsible for contacting private, wealthy individuals and families. I'd already identified many of these high-net-worth people through my previous positions, and through financial planning firms who offered energy as an investment product. I approached people that I knew would appreciate the value of this ground-floor opportunity.

I worked long, hard hours, seven days a week for that first couple of years, lining up leases and putting together investment pools. But although it was often difficult, I did it not only because I saw the potential, but also because I genuinely loved the thrill of the sale and exploring for oil and natural gas. This part of my nature may be why I get invited on trips to Las Vegas all the time, but I politely decline because I don't gamble in that way. Success and failure (failures more frequently than successes) are the nature of this business, and there is nothing more exciting and gratifying than making an oil or gas finding. It gets into a person's blood, and there is nothing else that I would rather do than search for the next big discovery.

Back in those days, before horizontal drilling techniques made oil recovery the booming business it is today, the only way to uncover oil was to dig vertical holes in areas that looked promising and cross your fingers that you found some oil. (OK, so maybe there was a *bit* more to it than that...) So although investors knew that it was a good time to invest, and were anxious to take advantage of tax write-offs, drilling a well was still a complex and fairly costly endeavor, particularly since it was hard to know whether a well would actually produce any oil. You might spend millions on all the expenses involved in drilling a well and extracting the oil, and only retrieve enough to break even (and sometimes, not even do that). The best-case scenario was when investors would put money into several wells—five or more, preferably—since the thinking was that the profitable holes would pay for the not-so-profitable ones. And that kind of investment was costly.

> The first successful oil drill in the United States was in Titusville, Pennsylvania in 1859.
>
> Source: "50 Surprising Facts You Never Knew About Oil," Investing/Answers.com

But the investment alone gave investors an initial write-off, reducing the costs right off the bat. (See Chapter 7 for more on tax benefits.) Plus, in addition to the IDCs that the depletion allowance provided them, if their wells were producing oil or gas, those investors could write off 15 percent of every dollar made. And even with a well that was a dry hole (a complete loss), investors could write off 100 percent of the loss.

So the smart investors bought in. Like master limited partnerships, which are publicly traded on the securities exchange, our deals paid dividends to investors. These dividends provided the monthly cash flow these individuals were seeking, so it was a win-win. In this way, we were able to put lease funds together and expand our drilling efforts.

Intangible Drilling Costs (IDCs) are costs that fuel the economy and create jobs—the equipment, labor, and other essential factors in drilling for oil and gas—which is why we are 100% deductible. In 2021, the United States oil and gas industry supported 11 million jobs and contributed nearly $1.8 trillion to the American economy, according to an independent study by PricewaterhouseCoopers, which was commissioned by the American Petroleum Institute.

Source: "USA Oil and Gas Supported Nearly 11MM Jobs,"
by Andreas Exarheas, May 19, 2023, Rigzone.com

Building REI

We started slowly, drilling primarily in northwest Texas. But by 1992, we were drilling in Southern Louisiana. And in the late '90s, we acquired our first international deal in Argentina, one of our most successful projects.

One of our geophysicists originally brought the property to us, saying it was believed to contain vast amounts of oil and natural gas. At that time, the Argentine peso was strong. The company that had owned the property had filed for bankruptcy and was hurting for cash. We bought it, then sold half to Argentine Oil Company, or Compañia General de Combustibles; the thinking was that a local company whose people knew the area would know more about the land and the intricacies of drilling in that country.

We ended up drilling over 20,000 feet, making the deepest discovery ever made in that country and the fourth-deepest well ever drilled in Argentina, and found the property to be a prolific source of natural gas.

But the operations there were costly and natural gas prices were cheap, and we became anxious to sell and move on. Fortunately, the American vice president of this Argentinian partner company came to us and said that a large public oil-and-gas company had expressed interest in purchasing the property, and we mutually decided that we would sell. However, not long after we'd made this decision, we were then told that this large public company had gone in another direction and would not be making the offer after all. The partner company offered to take the property off our hands; we sold our share for $55 million and called it a successful day.

That is, until the next day, when I opened the newspaper to find that our "partner" had turned around and sold REI's interest to Shell Oil for $175 million. We'd been double-crossed.

A lengthy legal battle ensued, which didn't conclude until late 2001, and although we won the judgment, we were unable to collect the lion's share of the money owed to us.

But we learned a valuable lesson about doing this kind of work and trusting others to call the shots. And since then, we've become one of the top independent oil-exploration companies in the United States. We've been a lot smarter about whom we partner with, when it's the right time to sell, and with what properties we get involved. It also helped me to truly understand how important oil is to the world, especially in developing areas.

So despite the struggles that came with being in the oil business during the Arab Oil Embargo and the subsequent price collapse, partnering with Vearl Sneed and forming REI Exploration in 1987 remains one of the best decisions I've ever made.

Since we formed the company, a tremendous number of changes have taken place in the industry—how and where oil and gas are used, how they are recovered, and the financial intricacies of investments made in these resources.

Which leads me to my next topic: Why do we need so much oil?

CHAPTER 3

WHY DO WE NEED SO MUCH OIL?

"The healthful balm, from Nature's secret spring,
The bloom of health, and life, to man will bring;
As from her depths the magic liquid flows,
To calm our sufferings, and assuage our woes"

— SENECA OIL ADVERTISEMENT, CIRCA 1850

By far, the United States is the #1 consumer of oil. In 2021, Americans consumed 19 million barrels of oil per day (BPD), with China at a distant 14 million BPD—although its population of 1.41 billion is more than four times that of the United States.

In the year 2020, despite the temporary slowdown in gasoline usage due to COVID-19, the production of oil and natural gas was still shockingly important. At an energy and manufacturing roundtable in New Mexico that took place in New Mexico, Mark Menezes, the

Deputy Secretary of Energy, stated, "Oil and natural gas provide more than two-thirds of the energy Americans consume daily. In addition to meeting our energy needs, these fossil fuel resources are integral to our standard of living."

Why do we use so much oil?

Often, we think of oil in terms of how much gasoline we put in our cars, forgetting that natural gas is not only how many of us cook, but it's also the largest contributor to the nation's power generation.

But that's only part of it. Every time you fire up your computer, paint your bedroom, put on makeup, wash the dishes, fertilize your lawn, have a medical procedure, use your credit card, or even pick up crayons to color with your children, you are using petroleum products. Of course, the largest percentage (19.45 gallons of a 42-gallon barrel) of crude oil used in the United States is in the form of finished motor gasoline, and about 17 gallons of diesel, jet fuel, and distillates.

We are a nation of drivers. In the third quarter of 2021, there were almost 284 million vehicles on U.S. roads. Proportionally, that's roughly one car for 85% of American residents. That figure represents a 10% increase over just four years previous. And while public transportation usage is increasing, we know it really only plays a major role in a few major metropolitan areas of the country.

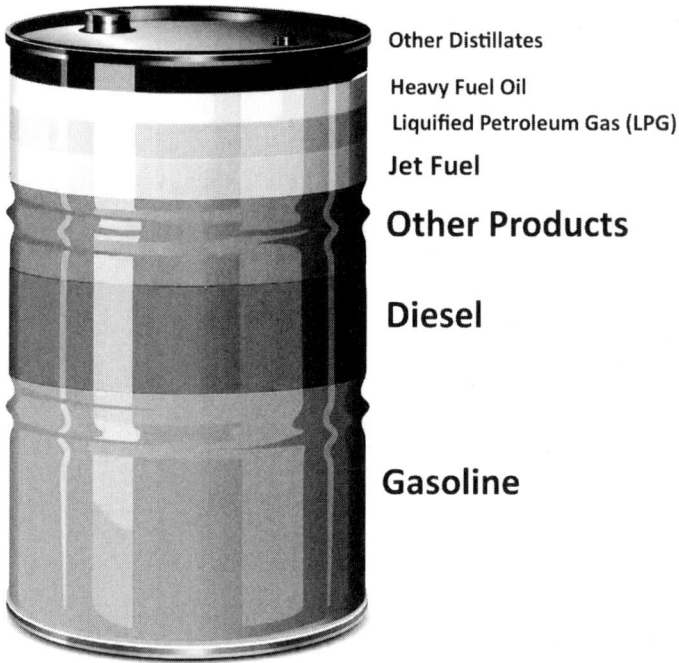

Other Distillates

Heavy Fuel Oil

Liquified Petroleum Gas (LPG)

Jet Fuel

Other Products

Diesel

Gasoline

Meanwhile, the remaining six or so gallons in each barrel of crude oil is comprised of other petroleum products. According to the EIA, here's just a sampling of other products that are derived from petroleum:

- Ink
- Candles
- Paint
- Cosmetics
- Safety glasses
- Crayons
- Dishwashing liquids
- Roof Shingles
- Fertilizer

- Tires
- Asphalt
- Antihistamines
- Trash bags
- Heart valves

Plastic is made from petroleum, and plastic is ubiquitous in our daily lives; our children's toys and car seats are made from it, and so are appliances, accessories, home furnishings, and more. In the health care industry, petroleum is used in the production of polymers, which make up such vital tools as syringes, plasma bags, tubing, prosthetics, pacemakers, or even certain medicines. Petroleum also is an essential component in paints, linoleum, asphalt, refrigeration, packing materials, and certain items used in the manufacture of airplanes. In short, nearly every industry and household is affected by and relies on petroleum.

Because oil is vital to the production of so many of the products we use every day, it's fair to say that oil will remain the world's most important commodity for the foreseeable future—with demand expected to continue rising through 2050, remaining the largest energy source, even ahead of surging renewable fuel use. This is because oil is economical to produce, we have an abundance of it here in the United States and new technology, including unconventional drilling methods (which I'll cover in Chapter 5), make it easier than ever to extract from the ground.

And it's because of this that oil should be a part of every investor's portfolio.

DID YOU KNOW?

Oil has been used for more than 5,000 years—the ancient Babylonians and Sumerians used crude oil as a medicine for treating ailments such as gout and frostbite.

Source: "50 Surprising Facts You Never Knew About Oil," InvestingAnswers.com

What about Renewable Energy?

We are asked all the time about the role renewable energy will play in people's future, and in the future of the oil-and-gas business.

Our response is this: Yes, renewable energy sources such as solar, wind, geothermal, hydropower, and biomass energy will be used with greater frequency year after year. This is because innovations increasingly make these sources easier and more economical to use. And, put simply, crude oil is a finite resource—not one that's likely to run out anytime soon, but it would be a falsehood to indicate that renewable energy is not needed.

"The difference in power generation between solar power and oil production is more than the difference between a professional bicyclist and a Formula 1 racecar."

Source: "What Are the Top Five Facts Everyone Should Know About Oil Exploration?" Forbes.com

However, at this point in time, renewable sources of energy are not practical or affordable enough to be easily tapped instead of oil and gas. That's why, in 2022, all renewables combined (solar, wind, geothermal, hydroelectric, and biomass) contributed only 13% to the total mix of U.S. energy consumption. The cost of the equipment alone will significantly impede the benefits these sources offer.

Part of the problem is the accessibility of the supplies needed to produce renewable energy—for example, the batteries needed to power electric vehicles. On a December 2022 episode of my podcast *The Energy Show with REI Energy*, I spoke with Henry Sanderson, a journalist and author of the book *Volt Rush: the Winners and Losers in the Race to Go Green*. Sanderson explained to me that the biggest beneficiary in the race for renewables is China. "China, at the moment, controls the lion's share of almost all clean energy technology manufacturing, from solar to batteries," he said.

Take a battery used to power an electric vehicle. The materials used to produce that battery include lithium, of which the largest supply in the world is found in Australia, followed by Chile. But the vast majority of the processing of that lithium is done in China. Meanwhile, there are minerals that are both mined and processed in China, including graphite and as rare-earth elements.

So the question becomes, do we import these materials—a costly and sometimes highly political endeavor—or do we start mining for them in the United States? And with that, you face NIMBY troubles: Not In My Back Yard. Any mine will have an impact on the environment, which means extensive regulatory hurdles to overcome. Any large-scale developer must issue an environmental impact statement first, and some of them can be over 1,000 pages long and take years to complete—time few developers can afford.

Plus, while the majority of Americans say they want to transition to new energy sources, they aren't often willing to sacrifice their views or current way of life to helping make that happen. Developers of wind turbines and solar farms, like mining companies, struggle mightily with NIMBY challenges.

"Anyone who spends time talking with renewable-energy developers knows that NIMBY-ism—people opposing new projects not in principle, but in their backyard—is a major barrier to building a clean-energy economy," reported Jerusalem Demsas in the October 2022 issue of *The Atlantic*. "And the permitting process creates ample opportunity for localized unhappiness to turn into legal or procedural barriers."

Sanderson says that although lithium battery development is picking up in the United States, with the growth of Tesla and others, supply chain challenges should be expected for the foreseeable future. "The United States has deposits of minerals. The problem is, can it actually start to mine them? The other thing is mines take a long time to come online. They can take a minimum of five years. So once you've gotten through the social and environmental hurdles, actually constructing the mine, bringing it to production takes time. So this decade, I think the West will have to rely a lot on imported minerals."

The Institute for Energy Research (IER) reported in 2021 that it would cost the United States *several trillion* dollars—about a quarter of the U.S. debt—to upgrade our power grid to a 100% renewable system. That translates to about $2,000 more per household each year until 2040. Clearly, today's costs far outweigh the benefits. Until the energy sources can be easily tapped for a significant return on investment, oil and gas will remain the top source of energy.

Renewable sources—large, open stretches of sun-drenched land, for example—are also usually located in remote areas, making it expensive to build the power lines and infrastructure needed to deliver this energy to cities. There's a reason why no power system in the world gets more than 30% of its power from wind and solar.

And, quite frankly, these sources are still dependent on the elements—it's hard to draw solar power on cloudy days or wind power on calm ones.

"In areas of the country that have a decent mix of wind and solar potential, those places can probably get to 50% renewables without struggling. Above 50%, the challenge of ensuring reliable grid operations starts to take off," said a spokesperson from Wood Mackenzie, a research firm conducting studies on the transition to renewables.

Finally, another problem with renewables is that, surprisingly, they aren't always clean. Take the example Sanderson uses of the carbon emissions coming from electric vehicles. One of those emissions is nickel, which is primarily mined in Indonesia, the world's largest producer of it. Processing that nickel and taking it through production relies heavily on coal-fired power, making it carbon intensive. Plus, only about 1% of the earth's minerals is comprised of nickel, meaning that extensive waste is created during nickel-mining efforts, not to mention high levels of pollution. While nickel can also be mined in Canada and Australia with a lower carbon footprint, it's not nearly enough to satisfy the growing demand for electric vehicles.

What's the upshot of all this? In 2022, renewable energy sources still only accounted for 13% of our nation's energy use, with non-renewables making up the remaining 87%. From this small set of renewable sources, biomass (wood, biofuels, etc.) comprised 37%. Wind made up 29%, hydroelectric totaled 18%, solar came to only 14.2%, and geothermal a measly 1.6%.

While I certainly realize the need to tap other sources of power and have no illusions that we can turn a blind eye to their advancement, the bottom line is that the infrastructure, supply chain, processes, and buy-in we would need in order to draw the majority of our power from renewable sources simply aren't available to us now, nor will they be anytime soon.

Clearly, non-renewable sources will remain our primary energy sources for years to come.

Also, it's important to realize that the concept of drilling for oil doesn't disappear when renewable energy begins making major strides. Even if every car on the road today were to be converted from gas to electric power, remember that just over 19 gallons of a 42-gallon barrel are used for gasoline. Oil would still be critical in the production of the plastic in their parts and their tires. We also will still need oil and natural gas products to power our homes and make all those products that we use and enjoy daily.

As Ryan Carlyle, a hydraulics engineer, says in *Forbes*, "To *replace* oil, you'll need a century to allow the entire economy to retool and realign around the new technology."

> "Three percent of today's vehicles are electric vehicles, but to get up to even 10%... all the lithium in the world combined is not enough to make batteries for more than a couple of percentage points increase in electric vehicle sales."
>
> Source: Jim Rickards, "Green Scam," The Energy Show with REI Energy podcast, 2019

Oil Consumption in India and China

Far more than renewable energy, world oil consumption is being influenced by two primary factors: India and China.

Each of the first and second most heavily populated nations in the world has a population in excess of a billion. As the third most heavily populated nation, the U.S. has less than one-third the number of citizens of both, and it still consumes far more oil than those two countries combined—the United States consumes 21% while China consumes 15% and India only 5%. American society is heavily reliant on gasoline and electricity, computers and wireless devices, and ours is one of the most modernized and developed in the world.

But while you can venture nearly anywhere in this country by car, and purchase gas or use your cell phone while you're there, the same can't be said of either China or India, which outside of their major cities are fairly, if not entirely, undeveloped.

There are still about 840 million people—which constitutes about one-ninth of the world's population—living in China's rural areas. These people include farmers, the elderly, children, and factory workers, and their villages often have no running water or electricity, and a year of agricultural work may help to eke out a salary of only 30,000 yuan a year—or just over 4,000 USD.

Nearly half the population of rural China resides in areas that still has no internet access.

And in India, the majority of the population—over 65% as of 2019—lived in what would be considered a rural area. Many rural residents farm, raise livestock, and fish. Millions in India still live without electricity. And India is one of the few places in the world where ox carts are frequently used for transporting goods; about 13 million of them still operate today.

However, things are changing rapidly. Agriculture has been in decline in rural areas, and increasing industrialization and technology— computers, cell phones, satellites—are driving huge waves of urban migration. This is expected to continue in the coming years, creating major economic shifts in these countries that will increase oil consumption. As Indian and Chinese citizens continue to demand more energy and more petroleum products, we'll see a massive sea change in the oil-and-gas business, with exports driving the industry here in the United States.

China became the largest importer of oil in the world in fall of 2013, surpassing the United States, according to the EIA. China is the world's largest car sales market, with 26.86 million vehicles sold in 2022, which was up 2.1% over the previous year. China's refusal to open its economy during the pandemic is seeing its economy falter, and the nation has put forth a plan to further boost car sales to help shore it up, meaning its fuel consumption will only increase.

Historically, the primary source of China's energy has been coal. In 2020, coal made up 56.8% of the nation's energy use. However, this is changing, and demand for natural gas, in particular, is on the rise in a major way. As of 2022, China is the world's third-largest gas market, increasing its domestic gas production by 10% each year for the last six years in a row. Clearly, natural gas production is on the rise around the world.

Meanwhile, the United States has become the world's largest crude oil exporter, thanks to modern drilling methods, which I'll be discussing later in this book. In fact, we exported more than four million barrels of oil per day in the first five months of 2023.

While China and India primarily import their crude oil and natural gas from Russia, the Russia-Ukraine war has dramatically shifted the landscape of the oil-and-gas industry. Despite Russia's influence on China and India, the number of U.S. exports of crude oil into Asia continues to grow. In fact, nearly half of U.S. crude exports go to Asia, to not only China and India but also Singapore, South Korea, and Taiwan. And with Russian oil being outright boycotted by much of the world in response to the war, the United States is the recipient of much of that business.

While our country is still the second-largest oil importer in the world, we're also on our way toward energy independence. And as other countries' dependence on our exports grows and costs rise as a result of global pressures, that's great news for investors in oil and gas.

CHAPTER 4

AMERICA'S OIL RENAISSANCE

"It is likely that the larger pools
have already been found..."

— CANADIAN GEOLOGICAL SURVEY 1975

When Vearl and I started REI all those years ago, we did so with the belief that what comes down must go up. We had seen and enjoyed an oil boom, and we'd dealt with the repercussions of a recessed economy and glut caused by overproduction here and around the world.

Fortunately, we were experienced enough to understand the boom-and-bust nature of the industry (we might not have started REI otherwise). We were confident in the knowledge that people would always need gas for their cars and petroleum in their plastics, so it was only a matter of time before the oil glut of the mid-'80s reversed itself. And because of this cycle, it wasn't too difficult to convince savvy investors that they would reap rewards by investing with us.

Now that more than 45 years have passed, we find ourselves in another American oil boom.

Bob Lyman, retired energy economist and former senior official with the Canadian government who advised on energy policy, explained on *The Energy Show* podcast in February 2023, over the last 20 years, global demand for oil has risen by almost 1 million barrels per day, per year, on average. "That's the fastest rate of growth in absolute terms of any comparable period ever," Lyman said. "That's in spite of the financial crisis that occurred in 2009–2010."

And it appears that rate of growth is continuing. It may seem odd, considering that demand from developed countries has largely been declining, but offsetting that decline is demand from the less-developed countries has risen by almost 9 million barrels per day, effectively wiping out any decline we might otherwise have seen.

To meet this escalating demand, which has been further heightened due to the supply chain issues brought on by the Russia-Ukraine war, U.S. oil producers are steadily pulling more and more oil out of the ground and in the third quarter of 2023 were on track to set a new record for output, with further growth expected in 2024.

All that growth is fueling employment, too. The Texas Oil & Gas Association reported that direct employment in the state's oil-and-gas industry rose 8.1% between 2022 and 2023, with corresponding wage increases.

So, how is this boom different from that boom of the '80s? Who's to say that an investment made today won't plummet as part of the cyclical bust that will inevitably come?

There are, admittedly, similarities between then and now. Political bickering was rampant in the late '70s and early '80s. Journalist Mycah Glover points out in a PermianBasin360.com article entitled "Clearing Up Oil Boom and Bust Speculation" that the United States was working with the Saudis back then, flooding the market with product to keep prices low and bring the Soviet Union to its knees. The Soviets had been a major oil producer and had gotten overly confident in America's dependence on what they had to offer. The Soviets embargoed all oil supplies to the United States. But when Saudi Arabia's resources became not only plentiful but far cheaper, the economy of the U.S.S.R. faltered, which many argue was the beginning of the end for the Soviet Union. The plan worked, but the U.S. economy paid a high price for it.

One of the only ways to put pressure on OPEC nations, such as Venezuela, Iraq, and Iran, is to flood the market with cheap oil. The majority of the revenue generated to support these countries monetarily is derived from the sale of oil. So when you want to get their attention, hit them in their pocketbooks. The tactical move to flood the market with oil can be devastating to these countries. So, of course, pricing matters are largely political.

OPEC countries formed an alliance with other oil-producing nations such as Russia and Mexico to form OPEC+ in 2016. Together, they produce about 59% of the world's oil. And to some degree, we're living through a similar period of geopolitical strife creating turbulence in oil supply and pricing, especially since the start of Russia's war on Ukraine in February 2022.

While the United States became the world's biggest oil producer several years ago, until recently, we were still importing millions of barrels of oil each day from around the world, including Russia. But when Russia attacked Ukraine, setting off a globally contentious war, all imports from Russia came to an abrupt halt through President Joe Biden's ban. To alleviate the burden of high gas prices and supply issues, he released our Strategic Reserves until we could begin producing enough of our own oil to address the demand.

But here's where things are different: The traditional boom-and-bust cycle that has historically led to overproduction has finally been broken. As a June 2023 NPR report by Camila Domonoske explains, pre-pandemic, the United States always would chase the money. When prices would increase, they would heavily ramp up production to take advantage of that price increase. One of the first rules of economics is that what goes up must come down. Eventually, all that production would lead to a glut of oil, causing prices to bottom out, resulting in a bust.

But these days, oil-and-gas producers are approaching things differently—in moderation. As Domonoske writes, "But a huge factor in this shift towards moderation is pressure from investors who want oil companies to share their profits with them, rather than funneling the earnings back into the ground to make more oil." She explains that investors are starting to demand a greater level of discipline from producers, which returns dividends and maintains a steady flow of oil production—not overproduction.

This restraint has helped producers to keep their heads nicely above water. Oil prices have remained high enough that most producers can drill new wells at a profit.

But aside from these things, there are notable, reassuring differences between the '80s and now, with the biggest difference being the technology that is used to capture these resources from the ground. It has increasingly become more effective and faster at finding and recovering oil, and it has become proportionately cheaper to do so.

In Chapter 5, I'll discuss the differences between conventional and unconventional drilling and the impacts these differences have had on the industry. But for now, I'll say that back in those days, there were a lot of dry holes, and a lot of drilling that led to little oil. Today's recovery methods are more accurate, and their resulting products more plentiful.

But another drastic difference between then and now is the proportion of foreign to domestic oil used in America and around the world. It used to be that, as the world's greatest consumer of oil, we were largely reliant on the Middle East—the most unstable region in the world—for most of our oil and gas. That situation has reversed in recent years, with technology making it possible to tap vast reserves of oil within our own country. While many Asian countries have increased their imports from the Middle East and Russia, here at home we're becoming ever more self-reliant.

So as Domonoske points out, "More discipline from American oil companies is also good for the global cartel known as OPEC+. The shale revolution has reshaped global oil politics, turning the United States into the world's top producer and an OPEC+ rival instead of just a customer. That means any time OPEC+ considers cutting production, it has to weigh whether U.S. producers will jump in to pump more crude, seizing more market share from the cartel."

Advanced technology has also allowed the United States to be a net exporter of natural gas since 2016, and the world's largest exporter of liquefied natural gas as of 2022.

Unlike the glut of the mid-'80s, when supply exceeded demand, we in the oil and gas industry are finding that as we increase supply, worldwide demand is also increasing, and it shows no signs of slowing anytime soon. That's why 16 new natural gas-fire power plants will begin commercial operations in 2023, and 20 more are slated for 2024 and 2025. All our newfound oil and gas resources have given us a great opportunity just in time to lessen our dependence on foreign oil. And we owe it all to shale plays.

Revitalizing the U.S. Shale Plays

It wasn't that long ago that shale was little more than the useless, sandy sedimentary rock layer that oil companies had to drill through to get to the "good" rock—sand, limestone, or dolomite, for instance—which contained reserves of oil. Shale shattered easily and was believed to contain little more than compressed mud and clay.

But thanks to the innovations of horizontal drilling and hydraulic fracturing that have been used since the 1940s but really only began to take off in a big way in the last 20 years, shale has become a vital source of oil that we previously had no way of tapping. Although crude was detectable in trace amounts in the shale, it was believed to be unrecoverable.

Large expanses of productive shale have been identified in many unassuming areas around the United States, and these "shale plays" are providing a bounty of crude oil reserves. Names like Bakken and

Three Forks, in North Dakota, are now on everyone's lips, thanks to reserves we've only relatively recently learned these areas possessed.

Other shale plays of major importance in the United States are the Permian Basin of West Texas and Texas-Oklahoma Panhandle, one of REI's primary areas of drilling focus; the Haynesville Shale of northwest Louisiana and East Texas; the Eagle Ford shale of South Texas; the Marcellus Formation, a largely untapped source of natural gas reserves that extends across the Appalachian Basin; the Utica shale of southeastern Ohio and West Virginia; the Granite Wash, a collection of plays in the; the Anadarko-Woodford play in West-Central Oklahoma; and the Niobrara shale play in the Rocky Mountains of Colorado, Wyoming, Kansas, and Nebraska.

For REI, the Bakken/Three Forks area is where we primarily set our sights. This formation is comprised of layers of black shale, siltstone, and sandstone that are rich in organic materials. Because the shale has marine origins—the deposits of the long-dried Williston Basin— this shale is also rich in oil and natural gas deposits. The Three Forks Formation, which is separate from the Bakken, lies below the Bakken at 10,600 feet below the surface. According to the latest USGS assessment, the two formations are estimated to hold 4.3 billion barrels of unconventional oil and 4.9 trillion cubic feet (tcf) of unconventional natural gas—more than double what was estimated in 2013, when the USGS last assessed the area.

Today, the Eagle Ford and Permian Basin plays in Texas, along with the Bakken/Three Forks area are the largest sources of tight oil, which refers to oil produced from low-permeability sandstones, carbonates, and shale formations. Together, the three areas produce over 8 million barrels of oil per day—accounting for almost all U.S. tight oil production.

Meanwhile, proved reserves of crude oil have increased significantly in the U.S. in the last decade. The EIA reported in early 2023 that proved reserves increased in each of the five areas with the most U.S. oil reserves in 2021, setting a new record: Texas, New Mexico, the Gulf of Mexico, Alaska, and North Dakota. This increase is almost entirely due to advancements in exploration and unconventional drilling methods.

"In 2021, the oil and natural gas industry directly provided 2.3 million jobs for American workers, paid $278.5 billion in wages and salaries and benefits and proprietors' income, and generated $773.6 billion in GDP. The industry's direct national impact on the US jobs, labor income, and value added ranged from 1.1 percent to 3.3 percent in 2021."

Source: PricewaterhouseCoopers LLP, "Impacts of the Oil and Natural Gas Industry on the US Economy in 2021," prepared for the American Petroleum Institute, April 2023, p. 6

Peak Oil

In 1956, a geophysicist named M. King Hubbert predicted that America's oil production would peak in 2000. He suggested that all the big oil fields had been discovered and there really were no more billion-barrel oil fields to find in this country.

As Hubbert pointed out, the way oil extraction works resembles a bell curve. After a drill is made, oil flows slowly at first, but then rapidly increases until it's flowing freely and heavily. But at some point, it reaches its peak and begins its slow, inevitable decline as the oil and gas are depleted and become harder to access. Eventually, they run out altogether.

At the top of this bell curve is what we call "peak oil."

As I'll discuss later in this book, with conventional drilling methods alone, Hubbert was correct. We reached our production peak in the United States in the 1970s, and were forced to turn more toward imports, from other countries that were producing far more than we could. United States production dropped considerably until 2008, until unconventional drilling methods became widely used and untapped reserves became something that, for the first time ever, we could reasonably expect to be able to tap. We began climbing another bell curve at that point.

But this concept of peak oil—a point at which the depleting fossil fuel resources would have reached their peak and would begin their inevitable decline—is a reasonable concern, and one that everyone in the industry knows we will face. Every day we have less oil—that's a simple fact.

But it's crucial to point out that, as Chris Martenson wrote in 2009 in his well-known book *Crash Course: The Unsustainable Future of Our Economy, Energy, And Environment,* "peak oil" is not the same thing as "running out of oil." Reaching our peak does not mean that we're about to run out. Actually, it's more like we have about half as much left. But at some point, it costs more to extract the oil than it's worth to retrieve it, and that's when a well is usually abandoned.

Martenson's book cautioned that we had reached our peak, and were on the precipice of inevitable decline, with no alternative sources on the horizon to replace fossil fuels in a practical way. He warned that although this didn't necessarily mean we were out of oil, it would soon lead to a supply-and-demand problem in which fear of running out of oil would cause hoarding and price gouging that would be outstripped by this fear, and chaos could ensue as a result.

With explosive population growth around the world expected to take us from today's 8.1 billion residents to roughly 10 billion in the next 30 years, and with Americans using more than 19 million barrels of oil per day, demand is skyrocketing. Peak oil is a genuine concern.

Martenson's book was published just as unconventional drilling methods were coming into their own, in a sense rescuing us from the fate he cautioned about.

However, once again, oil is a depleting resource. It's not self-sustaining, and although it does naturally regenerate, that won't happen during any of our lifetimes, or any time in the next million years. The minute you begin tapping an oil field, it's declining. And our whole world, as we discussed in the last chapter, runs on oil.

However, we believe we have enough oil reserves in the shale plays to last us for the next 100 years or so. Some say we haven't hit peak oil yet, with the number of reserves frequently increasing as the technology we use to extract oil constantly advances. We have drastically slowed the decline through unconventional drilling methods. At some point, there will have to be another primary energy source to replace hydrocarbons. That's an indisputable fact. Maybe this source will be solar or wind energy, or something no one has even discovered yet.

But right now, developing those other sources is simply too cost prohibitive. To compensate for our current oil use, each of these alternative sources would have to increase output by far more than is reasonable to expect.

Meanwhile, we have access to effective, relatively inexpensive energy, which we're pulling out of the ground safely and efficiently, thanks to continual innovation.

And that leads me to the subject of my next chapter: Exactly how *do* oil companies pull all this oil out of the ground, and what makes these shale plays so productive?

CHAPTER 5

DRILLING FORWARD: CONVENTIONAL VS. UNCONVENTIONAL DRILLING

"Prospecting for oil is a dynamic art...
The greatest single element in all prospecting,
past, present and future, is the man willing
to take a chance."

— TEXAS OILMAN, GEOPHYSICIST,
AND PHILANTHROPIST EVERETT DEGOLYER

The world—or, at least, the folks in the energy industry—had known since the 1940s that there were oil and natural gas reserves in shale, and that fracturing the rock would release those reserves. But as for how to get that oil and gas to flow in an economical way, in order to harness and use it? That was a mystery until 1998, when a Texas oil man named George Mitchell, spurred on by numerous geological reports showing vast reserves of oil and gas in the Barnett Shale underneath the Dallas-Fort Worth area, cracked the code for tapping those resources.

When Mitchell passed away in July 2013, he was considered the father of the modern American oil and gas renaissance. Shortly after Mitchell's death, *The Economist* memorialized him this way:

> From the 1970s America's energy industry reconciled itself to apparently inevitable decline. Analysts produced charts to show that its oil and gas were running out. The big oil firms globalized in order to survive. But Mr. Mitchell was convinced that immense reserves trapped in shale rock deep beneath the surface could be freed. He spent decades perfecting techniques for unlocking them: injecting high-pressure fluids into the ground to fracture the rock and create pathways for the trapped oil and gas (fracking) and drilling down and then sideways to increase each well's yield (horizontal drilling).

Getting Oil from a Sponge

Think of shale, which is mostly made up of compressed sand and clay, as a sponge. And like a sponge, it's made up of cells, or pockets, that contain bits of oil and natural gas. But imagine trying to pull the water out of any of the individual cells of a sponge.

It's darn near impossible unless you squeeze it—which you obviously can't do with shale.

Decades ago, engineers figured out that they could fracture (or "frac") the cells to release the reserves inside, but, like trying to capture a drop of water from one sponge cell after squeezing the entire sponge, capturing the methane once it was released from the cell was the challenge that had stumped so many in the industry. Getting it to flow and then being able to retrieve and use it was the key.

This is what George Mitchell figured out. His company, Mitchell Energy & Development Corp., invested roughly $6 million over a 10-year period starting in the late 1980s, in the effort to make fracturing a viable means of extraction. They had hit upon the idea of forcing drilling fluids through the fractured cells to release the reserves, and although it worked, the costliness of the method and the materials involved meant that the end didn't justify the means.

But Mitchell's gut told him that there was something to it, if he just kept working at it. He told *The Economist* in a rare media interview in 2012, "I never considered giving up, even when everyone was saying, 'George, you're wasting your money.'"

Mitchell was almost 80 years old when his team finally figured out that forcing sand and water—not the thick, gloopy drilling fluids they had been using—through the cells got the gas flowing at a reasonable cost. This meant that hydraulic fracturing was now a cost-effective method of harnessing those extensive, untapped reserves.

But when it comes to unconventional drilling methods and fracturing shale, it's about gathering all the reserves from the cells in that enormous shale sponge, which spreads horizontally for miles under the surface. In order to cover the largest area in the least disruptive way possible, you need to drill horizontally. In 2002, Mitchell helped to turn this development into a boom by implementing horizontal drilling practices.

In the past, with conventional drilling, oil companies hired geologists and geophysicists to collect seismic data and explore for oil and gas. Once they came back with data showing the presence of these resources, companies would go out and bore vertical holes into the source rock containing oil or gas. Typically, the hit ratio on this kind of exploratory drilling was about one in 10; we'd hit nine dry holes before producing a good well. You had to find the pool of oil, and when it was dry, you might keep poking holes unsuccessfully in order to find another, which is expensive and, as you can imagine, very disruptive to the surface.

But it was also very costly for these companies. In the 1950s, it was a lot easier to hit oil because, put simply, there was a lot of undiscovered oil territory. But before long, it became more difficult as the more obvious reservoirs had been tapped. And you had to drill several wells until you found one that could help you make your investment back. By the 1970s, a lot of major U.S. companies left the states because they'd already found most of the conventional reservoirs, and the areas in the Middle East held much more promise.

When those major companies left, it left room for smaller independents, like REI, to explore for conventional reservoirs, and these companies could take greater risks for exploration. At the helm of another independent, George Mitchell took a risk on horizontal drilling, and it paid off.

Mitchell owned a natural gas pipeline that ran from Texas, where his conventional wells were producing oil and natural gas, to Chicago, a major hub for distributing the oil and gas to the United States. But for the offshore rigs, conventional wells were difficult, because the ground beneath the water was often uneven, and he didn't have the luxury of being able to put a drill directly over the reservoirs of oil. For those, he had to put in a stationary platform, set the pilings into the seabed, and drill wells off of it. He had to find a way to send drills into different directions, to go where the reservoirs were. That's how his company ended up developing horizontal drilling, which eventually became the most effective way of pulling reserves from shale.

Put simply, horizontal drilling involves drilling vertically, and instead of stopping there, the well kicks off horizontally. That point at which the well makes a 90-degree turn is called the "heel" of the well. Then the pipe stretches out for roughly the same distance as it was sent vertically. The end of the horizontal section is called the "toe." As you can see, twice the area of pipe is going to produce far more oil.

By 2008, horizontal drilling was in widespread use by a number of oil-and-gas companies, and when the technology had advanced to the desired level of efficiency and affordability, we at REI began using the unconventional methods of fracking and horizontal drilling, which we now do almost exclusively.

The beauty of these unconventional methods is that because we are plunging into these wide expanses of shale, we always get a well, every time.

"From Exxon Valdez to Deepwater Horizon, these are the kinds of images we've come to associate with oil in the ocean. But a highway or parking lot full of cars could be an even more accurate—if less charismatic—symbol. Land-based runoff is the top source of oil to the sea and up to 20 times higher than it was 20 years ago.... The runoff flows from cities, highways and vehicles to rivers and the ocean....

The second biggest source is natural seeps, which is when oil enters the ocean through fractures or faults in the seafloor."

Source: "Most Oil in the Sea Comes from Runoff on Land," by Kat Kerlin, Oct. 3, 2022, UC Davis

Now, do we always make enough money and find enough of an accumulation of oil and gas to pay for the cost of the well? Not necessarily. But the odds are always in our favor now. Let me explain.

Let's take one square mile of land, or 640 acres, as an example. With conventional drilling, I would drill down 8,000 to 10,000 feet for one wellbore, and because I'm only allowed by law to drill every 40 acres, I can only produce the oil from that 40-acre radius around the well bore. Within a 640-acre space, I would have to drill 16 wells to recover all the oil contained there.

But with most unconventional wells, my wellbore would still be 8,000-10,000 feet, but then the well would take a horizontal turn, and reach out for perhaps up to 10,000 feet. So I can recover all the oil from that entire area with just one well. With current technology, on that same 640 acres, I would only need four wells to recover the greatest amount of oil if I'm drilling horizontally.

It's not hard to see how cost-prohibitive conventional drilling is. With the cost of one conventional well sitting at around $6 million, the 16 wells I'd need for my 640-acre plot of land would cost me a total of $96 million to drill. Although unconventional wells are more costly, at $10 million per well, I only need four of them to drain the oil from this plot of land—likely recovering more from the shale than I would through conventional methods—and it would only cost me a total of $40 million to do so. The initial investment is much less expensive, and the return on my investment is significantly higher.

It's no wonder the majority of oil companies are drilling unconventionally. In 2008, only one in 10 active drills were unconventional, totaling just 28,409 wells. Just one decade later, in 2018, that number had increased to nearly 140,000 horizontal wells.

That's not to say that the U.S. oil-and-gas industry has entirely abandoned conventional wells, but unconventional drilling is no longer the exception—it's the rule. Stanford University reported that the number of unconventional, non-vertical rigs increased from 28% in 2010 to 80% in 2020. Meanwhile, globally, the vast majority of oil (92%) and natural gas (79%) production is still done via vertical wells.

Nonetheless, because of the decreased costs of unconventional drilling and the increased access to reserves, we are experiencing an oil and natural gas boom that shows no sign of slowing in the near future.

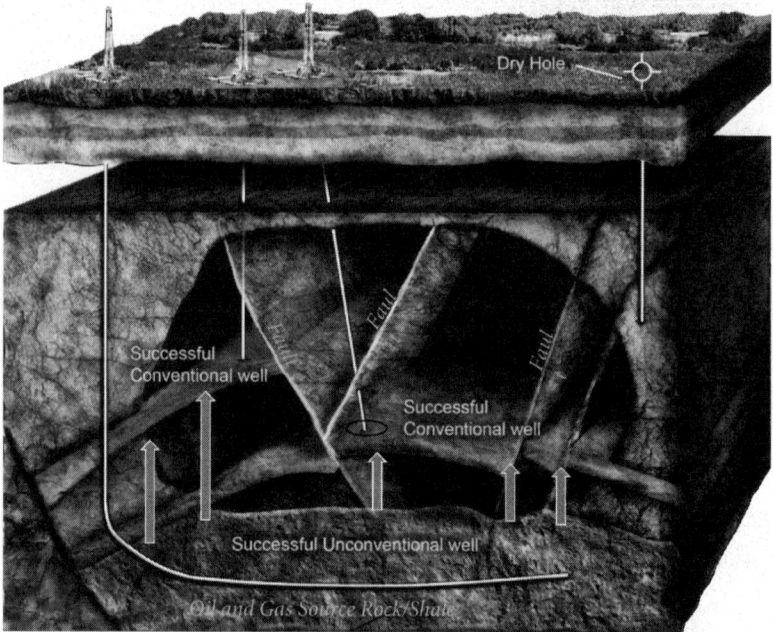

The Fracas about Fracking

Unfortunately, despite what fracking has done for our economy and for America's developing energy independence, the majority of questions we at REI receive have to do with the environmental impacts—specifically, the misconception that all fracking is destructive to the environment. Secondarily, many argue that long-term exposure to fracking sites can cause health problems.

How It's Done

Technology has significantly changed the way fracking is done. As I've said before, fracking isn't new—it's been around since the 1940s. The process really only needs two things: 1) enough pressure to break up rock, and 2) a fluid, which you inject into the rock through high pressure, thereby forcing the oil out.

Typically, this fluid is composed mostly of natural elements: water and sand. Anywhere from 95 to 98% of it is water. The rest is a combination of sand and a chemical, called a carrying agent, which is named for its ability to carry oil out of rock, enabling us to capture and use it. Only about 2% of any fracking fluid is this carrying agent.

To perform hydraulic fracturing, a large truck containing this fluid is positioned at the surface of the well. A pump on the truck is attached to the pipeline drilled into the ground. The truck begins pumping fluid at high pressure into the surrounding shale, with the intent of cracking it, enabling the oil and gas locked inside to flow into the pipe.

Fracking is done in stages, with a fracture at the well's heel being the first, and the rest occurring progressively down the pipeline. The oil then flows up to the well head for recovery. When this process was first developed, there were really only 10 stages. It was difficult in those days to get enough energy into the ground to fracture at the toe of the well. Most of the energy was in the first stage, and they only retrieved a limited amount of oil and gas.

But as technology has evolved, swellable packers have been developed. These long, rod-shaped tools are inserted into the wellbore and swell when they come into contact with fracturing fluids. By inserting these packers at points throughout the well, oil and gas may be pushed all the way to the toe. Generally, they're placed 100 to 200 feet apart, with about 40 stages of fracking per well. So now there's as much energy at the toe as there is at the heel.

The water breaks up the rock, and the carrying agent allows the oil to flow. But what does the sand do?

When you inject that much pressure into the earth, it wants to close in on itself. The sand is the most preferred propping agent, due to its uniform size and shape and its crush resistance; the sand keeps the cracks in the shale open while allowing the permeability that we need to get the oil to flow.

This is something else technology has changed—some companies have turned to using man-made ceramics as a propping agent. This is because sand eventually collapses, whereas ceramics are more secure and will prop up the ground for a longer period of time, keeping that shale flowing to allow us to retrieve more oil and gas.

The Controversy

Fracking has been widely used, and safely so, for about 70 years. Opponents of fracking will say it's bad for the environment, and many of these opponents are very outspoken about it. Yes, when not performed responsibly, fracking can be very detrimental to the environment. Fortunately, we're in a golden age in which technological advances have made it possible to extract the maximum amount of crude oil with a minimal amount of impact.

I want to address some of the biggest concerns people have about fracking, and how the oil-and-gas industry has made incredible strides in extracting oil through hydraulic fracturing, helping to lower electricity rates, stabilize gas pricing, and lessen the environmental impacts for an overall net benefit.

Water Resources

The biggest worry is that fracking uses such huge amounts of water, which must be transported to the fracking sites, that it depletes valuable resources.

There is no question that this process uses a lot of water. One well may use up to 20 million gallons of water to drill and frac a single well, with perhaps 40 separate fracking operations along the horizontal part of the well.

However, not all of it is freshwater drawn from groundwater sources. In fact, oil-and-gas companies are increasingly looking at recycling the water used to fracture the shale and generate oil, which is left behind after the oil is tapped. After all, managing the wastewater created through fracking is a major concern as well. Reusing it not only minimizes the impact on freshwater sources but also provides a way to use wastewater, rather than disposing of it. In the August 2022 *New Horizons: Hydraulic Fracturing Techbook* published by Hart Energy, an article by Paul Wiseman reported that 10 to 20% of produced water is being repurposed for fracturing, with some producers doing as much as 50%.

Plus, recycled water might actually be better for oil extraction. "'It's been established for a number of years now that, if anything, using recycled water in fracs generates better results than using fresh water,'" said Josh Adler, founder and CEO of Sourcenergy, which Wiseman explains is "because water from the same formation is more compatible for the frac."

Issues with the supply chain of slickwater and sand for fracking along with legislative and public pressure to responsibly manage water resources means that oil producers will increasingly be recycling their water with every passing year.

Chemicals

Another issue many people raise is that the carrying agents in the fracking fluid are potentially carcinogenic chemicals, and that they may contaminate groundwater once injected into the fracking site.

It's important to distinguish here that it's bad practices by irresponsible companies that would cause this to happen—not fracking itself, which is a highly regulated activity. The only evidence of such leaking has occurred with faulty wells.

Oil-and-gas producers are subject to the Safe Drinking Water Act, so the rules around what can go in and around drinking water reserves are quite strict—with good reason. Again, fracking is 95 to 98% fresh water, while only about 2% is carrying agent, sometimes called slickwater. So very little of this chemical is going into the ground.

Additionally, most wells and drinking water sources are only hundreds of feet underground, while we're usually fracking at 10,000 feet. So the fresh water is protected by the string of pipe reaching down to the bottom of hole at 10,000 feet. The fracking fluid is cemented in; unless you have a bad pipe and aren't properly monitoring your activities at the site, you simply won't get chemicals or fracking materials into fresh water. So in the shale plays, which are heavily regulated, contamination of freshwater zones just isn't easy.

Additionally, there has been considerable innovation in the industry where chemical usage is concerned. For example, Locus Bio-Energy

Solutions is a company that manufactures biosurfactants. Surfactants are chemical compounds that decrease surface tension between two liquids and act as emulsifying agents; hydraulic fracturing often involve surfactants in enhanced oil recovery. Biosurfactants, like those made by Locus, are produced through fermentation and involve zero-carbon chemistries. They are fully biodegradable, sustainable, and can boost production in a single shale well. They also are required in lower dosages.

Earthquakes

Another worry when it comes to fracking is that forcing such pressure into the earth can cause earthquakes. As you pull oil out or dispose of used water inside the earth, you can get some small earth tremors.

"It's always recognized as a potential hazard of the technique," said Professor Ernie Rutter from the University of Manchester, in a BBC article. "But they're unlikely to be felt by many people and very unlikely to cause any damage."

Research by the Center for Integrated Seismicity Research at the Bureau of Economic Geology found that most earthquakes are associated with disposal in deep zones. As a result, many operators have voluntarily reduced their deep disposals into zones associated with seismic activity. And a team of MIT researchers have developed a method to manage human-induced seismic activity, simply by adjusting rates of fluid injection.

Bottom line, with energy and high pressure going into the ground at 10,000 feet or more, packing the earth with sand and shale, you are likely to feel tremors, but you are not damaging the reservoirs, and you will not collapse the hard rock in the earth, causing giant earthquakes. Typically, the seismic energy involved in fracking measures a magnitude of 2.0 or less.

Greenhouse Gas Emissions

Considerable environmental benefit is coming from fracking, which drastically decreases our footprint on the land's surface, since we don't have to drill as many wells to pull oil out of the ground as we once did. It only takes one rig to drill four wells all at the same time. Because we are able to draw more abundant, less expensive gas supplies, we've also reduced our reliance on coal and decommissioned coal-fired plants.

In fact, between 2005 and 2020 carbon dioxide (CO_2) emissions were reduced by 970 million tons per year as a result of fracking, while in China, 4,689 annual million tons were added, thanks to that country's heavy reliance on conventional drilling methods.

Advances in each frac site, including longer horizontal wells and simul-fracturing techniques—in which two wells are stimulated simultaneously, enabling more efficient oil production—help to limit the carbon footprint at drill sites as well.

Fracking operators also are cutting both costs and emissions. One huge source of emissions is truck traffic, as trucks are needed to transport water and sand to the well site. Investments in frac water pipeline systems and locally sourced sand reduce truck mileage.

Another shift these companies are making is from frac fleets powered by diesel fuel to those that are electric- or natural-gas-powered. Many operators have replaced their diesel pumps with dual-fuel or dynamic-gas-blending versions, which use natural gas to offer partial diesel displacement.

Moreover, electric submersible pumps now produce greater lift capabilities, so fewer pumps are actually required. "Wells that previously required five or six pumps now need just three or four to achieve the same amount of life," said David Baillargeon, engineering supervisor for ChampionX's Unbridled ESP Systems, in an article published by Hart Energy. "There's a benefit to that: a 1-ton reduction in carbon emissions and 20% fewer heavy metals per ESP install."

It's also worth pointing out that oil-and-gas companies take environmental protection laws and concerns seriously. Our industry is on the forefront of innovation in this area. For example, ExxonMobil Corp. ranked among JUST Capital's 32 Industry Leaders for Environmental Performance in 2022, for its ongoing biodiversity assessments and high level of carbon capture and storage.

Meanwhile, Honeywell has taken on management of methane, which is a more potent greenhouse gas than CO_2 and is a genuine concern for those opposed to fracking. One step Honeywell has taken is Gas Cloud imaging, which provides continuous, near-real-time, video-based leak detection and quantification.

Methane emissions have generally been declining in recent years, thanks to technological innovations. While natural gas production increased by 95% and oil production by 54% in the period between 1990 and 2020, methane emissions declined by over 15%. Many fracking opponents mistakenly believe that the oil-and-gas industry is the primary culprit in methane emissions—it's not. The number-one source of methane emissions is livestock. The U.S. oil-and-gas industry accounts for only 1.4% of worldwide methane emissions.

A Washington, DC-based consortium of the top 100 oil-and-gas companies in Texas also pooled their knowledge and resources to tackle methane emissions in the industry, proactively finding and fixing emissions to cut them by 14%.

The EPA also administers Clean Air Act regulations for the production of oil and gas, which includes rules around reporting greenhouse gas emissions. The agency, along with the Natural Gas STAR program and other partner companies, has identified and promoted several new technologies and practices that can reduce methane emissions in a cost-effective way.

"According to a U.S. Chamber of Commerce report, halting hydraulic fracturing would eliminate 19 million jobs (direct and indirect) between now and 2025."

Source: U.S. Department of Energy, "The Economic Benefits of Oil & Gas," October 2020

And let's not forget that fracking and horizontal drilling have led to more jobs, lower energy costs, huge profits for investors, and economic benefits for communities when landowners receive generous royalty payments. Not to mention the benefits we enjoy from simply having plentiful energy and petroleum resources at our disposal.

In the next chapter, I'll talk more about the significant economic benefits associated with the oil-and-gas industry.

CHAPTER 6

THE TAX BENEFITS OF INVESTING IN OIL AND GAS

"Profitable exploration requires wise investment of risk capital in people's ideas."

— AUTHOR MORGAN DOWNEY, OIL 101

So far, I've shared with you my own experiences working in the oil-and-gas business. I've described what's happened in the industry over the last several decades. I've explained why the business is booming, and why we can expect this to continue for some time to come. And now I want to show you why making an investment in oil and gas can be one of the smartest things you can do with your money.

As I stated in my introduction to this book, investing in oil and gas is a lot like investing in real estate—it is a far safer investment than the stock market will ever be, consistently throwing cash flow your way. By now, you can plainly see why: People will always need roofs over their heads, heat in their homes, and petroleum to make nearly everything they use in their daily lives.

Among all the investments you can make, oil and gas are perhaps the most beneficial, in terms of the tax incentives offered to both investors and small producers. Investments in real estate, the stock market, and retirement accounts, by comparison, are more limited in terms of their deduction and earnings potential.

The U.S. tax code incentivizes oil-and-gas production and investments because it plays a crucial role in meeting our energy needs. By diversifying your portfolio with oil and gas, you have leverage to protect against the impact of rising energy prices, you become less dependent on interest rates and the economy, and, with a strong worldwide market for oil and natural gas, along with the promise of energy growth in the United States and around the world, you have potential for good returns with regular monthly cash flow.

The tax advantages for oil-and-gas investments are substantial, from major deductions for tangible and intangible drilling costs as well as operating and administrative costs to tax exemptions for certain types of investment, and other incentives. Such advantages not only are available at the federal level, but there are certain state and local tax benefits as well.

As Tom Wheelwright, best-selling author of *Tax-Free Wealth* and Robert Kiyosaki's personal tax advisor, points out, oil and gas together make one of the greatest tax shelters available to Americans. To shed light on the various tax and real estate benefits available to investors in oil and gas, I've asked Tom to describe these benefits in detail and share how they work.

Tom has generously agreed to sharing some of his insights from his book *Tax-Free Wealth*...

Oil: We love to use it and hate to buy it. It has thousands of uses, from gasoline to plastic bottles to medical devices. It is messy, can ruin our clothes, and leaves nasty stains in our driveway. Truly we have a love/hate relationship with oil.

Personally, I love oil because I invest in oil. Why? Among other reasons, it tends to produce income for a very long period of time (a single well can easily last 30 years), and the price of oil just keeps going up and up as more countries develop and need gasoline for cars and heating oil for furnaces. One of my favorite reasons to invest in oil, though, is the amazing tax benefits I gain from it.

Numerous oil and gas advantages are allowed, including the following:

- Intangible Drilling Costs (IDCs)
- Tangible Drilling Costs
- Lease Costs
- Intangible Completion Cost
- Depreciation
- Depletion Allowance

Oil and Gas Tax Benefits

The United States has long held an energy policy promoting oil and gas drilling operations inside the borders of the United States to help reduce dependence on foreign oil. The government has put this into action through the tax law by providing significant tax benefits for anyone who invests in domestic (U.S.-based) oil and gas drilling operations.

Here's why oil and gas are one of the truly great tax shelters in the U.S.: Unlike real estate and other passive investments, oil and gas deductions are not subject to the restrictions on using passive losses. In fact, an investment in oil and gas is the only one not subject to these rules. That's right—*if you invest wisely, you can deduct losses from oil and gas against ordinary income, even if your investment is entirely passive.*

There are four types of investment in oil and gas. The first two don't really offer any particular tax benefits. The first type of investment is buying stock in an oil and gas company. This is treated like any other stock investment and has no special rules or benefits. Second, you can buy an interest in the royalties from a producing oil and gas well. This income is portfolio income, and other than creating investment income so you can deduct investment interest expense, there are no great tax benefits from investing in a royalty interest.

The other two types of investments in oil and gas are both investments in the actual drilling operation, and they provide great tax benefits. You can either invest in exploratory operations, also called "wildcat" drilling, or you can invest in development operations. Exploratory operations can be very risky, as there is no assurance that there is oil in the ground where you are drilling. Of course, with ever-evolving technology, this risk is always decreasing with the better operators.

Development wells are drilled in established oil fields where the reserves of oil are proven. The developer may need more money to drill additional wells, in order to get more oil out of the ground.

This tends to be less risky than the exploratory drilling, though no investment in oil and gas is guaranteed 100 percent—you can still lose your money. I once invested in a development well where the developer had absolute proof of huge reserves. The only problem was that we couldn't get to the oil. So, we lost our investment.

FOUR TYPES OF OIL AND GAS INVESTMENTS		
	TYPE	TAX BENEFITS
1.	Buy stock in an oil and gas company.	No
2.	Buy an interest in the royalties from a producing oil and gas well.	No
3.	Invest in exploratory operations, also called "wildcat" drilling.	Yes
4.	Invest in development operations.	Yes

"Oil and gas tax benefits" section excerpted from Tom Wheelwright's *Tax-Free Wealth*©. Reprinted with permission from the author.

Drilling Costs

When a development company drills for oil and gas, it has two main categories of expense. The first of these is Tangible Drilling Costs. These costs come from the equipment it purchases to drill, and they usually comprise about 20-40 percent of the cost of drilling a well.

The second is intangible drilling costs. IDC deductions have been allowed in the U.S. since 1913, to stimulate investments in the high-risk industry of oil and gas. These are the costs that are involved in drilling wells—things that go beyond the actual costs of purchasing equipment, such as casings and pumps. IDCs include such things as labor, survey work, ground clearing, drainage, fuel, and repairs. They generally constitute 60-80 percent of the total cost of drilling a well.

IDC expenses normally would be included in the cost of the well and would be depreciated or amortized over the life of the well. However, Congress decided to allow people to deduct their IDCs in the year they spend the money, which is usually the first year or two of investment in the drilling operation. That means that 70 percent of your investment is typically allowed as a deduction in the year that you make your investment in the drilling operation or in the following year. So, if you invest $100,000, you get a deduction for $70,000 almost immediately. At a 40 percent tax rate, that's the equivalent of the government giving you $28,000 ($70,000 x 40%) for investing in an oil and gas operation. Plus, you get to take depreciation on the equipment over the next several years.

Lease Costs

Lease costs include the cost of purchasing lease and mineral rights, lease operating costs, and all administrative, legal, and accounting expenses. These are all 100 percent deductible in the year in which they're incurred.

Intangible Completion Costs

Similar to IDCs, intangible completion costs are the costs of completing a well that are not physically recoverable—including labor, completion materials, fluids, etc. These things generally account to about 15 percent of the total cost of completion and are deductible in the year in which they are incurred. Hydraulic fracturing is considered a substantial intangible completion cost.

Depreciation and Depletion

The salvageable equipment used in the completion and production of a well is depreciated over a 7-year period, much as you would depreciate assets in any other industry.

However, you also get to deduct 15 percent of the well's gross income each year. This is called depletion.

The IRS defines depletion as "the using up of natural resources by mining, quarrying, drilling, or felling." In other words, because resources such as oil and gas are depleted as they're extracted from the ground, the IRS allows for a reasonable income tax deduction based on that depletion of the resources.

It's like depreciation, only you get it every year, even after you have deducted all of the IDCs and the depreciation. Basically, it's a gift from the government. Gross income includes all of the sales proceeds from the oil and gas and isn't reduced by any expenses. So, you could have $1,000 of gross income, and expenses of $400, for a net income of $600. You would then get a depletion deduction of $150 ($1,000 x 15%) and only pay tax on $450 of income ($600 less $150 depletion).

TAX BENEFITS FOR INVESTING IN OIL AND GAS	
1.	Deduction of intangible drilling costs, usually the first year of investment of drilling operations.
2.	Deduct 15% of the gross income from the well each year (called depletion).

In order to qualify for IDC and depletion deductions, you have to own a direct interest in the drilling operations. Owning stock in the drilling company or owning a royalty interest in the oil and gas doesn't qualify. Be sure to meet with your tax advisor about this *before* you invest in oil and gas.

And one other thing: In order to get all of the IDC deductions to which you are entitled, you have to own your investment through a general partnership or sole proprietorship. You can't own it through a corporation, limited liability company (LLC), or limited partnership (LP). If you live outside the U.S., be sure to check on your country's tax laws to find out what tax benefits you're allowed.

Additional Benefits

These aren't the only tax benefits associated with oil and gas investment. You can also deduct certain expenses incurred for the day-to-day operations of wells and earn credits for that enhance oil or gas production from a well. There are others, too, which is why it's important that you speak to a knowledgeable tax advisor about your investment portfolio.

Avoid the Passive Loss Rules on Oil and Gas Investments

In my book, *Tax-Free Wealth*, I explain that when you don't actively participate in a business, the income and losses from that business are treated as passive income and passive losses. When you invest in oil and gas development, you probably won't actively participate. And in the first year, you'll share in a tax loss of 70 -80 percent of your original investment.

So, wouldn't these losses be passive? Wouldn't they only offset passive income? Actually, no.

There is an exception to the rules about passive loss when it comes to oil and gas. Even though you don't actively participate in the business, losses from oil and gas are treated as ordinary losses, and you can take 100 percent of your losses against any kind of income. But in order for your losses to be ordinary, you have to be very careful about how you own your oil and gas investment.

You have to own your interest in the oil and gas development outright. You cannot own it through an LLC, LP, or any other entity that limits your liability. Instead, you have to rely on the developer's insurance to protect you from any lawsuits or disasters. Usually, the developer will form a partnership for the investors, and everyone will be a general partner for the first year or two.

In the second or third year that you own the investment, when the well begins to produce taxable income, you can change your ownership to a limited partnership or LLC to protect you from future liability. Good oil and gas developers will automatically change your ownership from general partner to limited partner as soon as the investment begins earning income (one reason why finding the right oil and gas company is so important). So, your liability is only for the year or two that the developer is actually drilling the well. [Excerpt from *Tax-Free Wealth*. Reprinted with Permission.]

For more from Tom Wheelwright on how to permanently reduce your taxes through businesses and investments, go to www.taxfreewealthbook.com

Now that you've gotten a primer on the tax benefits of oil-and-gas investment, let's take a look at an example.

Let's look at an actual investor who invested $100,000 in the REI Energy Drilling & Income Fund II, L.P. You know that intangible drilling costs (IDCs) represent the largest tax deduction you can take for this kind of investment; up to 70% of your investment can be written off for the year in which the wells are drilled. And the depletion allowance accounts for the reduction of reserves as the resulting product is produced and sold; after all, oil and gas are depleting resources. Up to 15% of the production revenue received from investments in oil and gas is tax free.

Here's the formula you would follow:

> $100,000 investment x 70% (IDC deduction) x tax bracket percentage = real dollar savings

So if you multiply that investment by 70%, you get $70,000 that you can deduct. If you multiply that by, for example, a 35% tax bracket, your real dollar savings is $24,500, which represents a 24.5% return on investment, just from your tax benefits alone.

That doesn't even include the income generated from such investments. In terms of the REI II investment fund I referred to above, as of July 2023, REI II provided cash distributions of $33,799. Add this to $24,500 in estimated tax savings, and you have $58,299 as a total realized return of capital.

Meanwhile, the depletion allowance on net income, in this example, would be 15% of $33,799, or $5,059.85 in tax-free income. And REI II has averaged 21.35% cash flow for 19 months. The average monthly distribution is $1,778.

CHAPTER 7

INVESTING IN THE FUTURE: WHAT YOU NEED TO KNOW

"We usually find gas in new places with old ideas. Sometimes, also, we find gas in an old place with a new idea, but we seldom find much gas in an old place with an old idea. Several times in the past we have thought that we were running out of gas, whereas actually we were only running out of ideas."

— GEOLOGIST PARKE A. DICKEY

The three most important qualities to look for in any property have always been, "Location, location, and location." This is especially true when it comes to oil-and-gas investing, since a failure to understand your location can land you in a dry hole or non-commercial well—literally.

Unless you have a line on a conventional well with a proven pool of oil just waiting for an investor, know that most unconventional plays cover enormous amounts of acreage, and while you'll always find oil this way, not every single acre will actually produce an economical amount of oil worth drilling. For example, the Bakken, one of REI's main areas of focus, runs 250 miles north to south, and 150 miles east to west. According to the USGS in 2021, there are about 4.3 billion barrels of oil and 4.9 trillion cubic feet of natural gas in recoverable reserves in the Bakken and Three Forks formations, but they aren't evenly spaced around the region. The key is knowing where to be in that shale play, because some spots are better than others.

That's why we employ experts. Our executive teams of land professionals, engineers and geologists are constantly tracking the producing wells in that basin, looking for those areas where we have the potential to recover at least 100,000 barrels of oil in our first four to six months; the highest-volume recovery period is those early months, with production slowly rising to a peak and then tapering off. It's only worth our while to lease properties in areas that will produce at least 400,000 to over 1,000,000 total barrels of oil for us.

So, much like you would when choosing a new home for your family, we're shopping for "good neighborhoods" in the Bakken and are careful to avoid the less-desirable ones.

The Bakken shale formation is located in the Williston Basin, which is, for all intents and purposes, a giant bowl comprised of the porous, semi-permeable shale rock I likened previously to a sponge. Almost dead center is its deepest point, plunging to about 16,000 feet below the surface, underneath Williston, North Dakota. The basin thins out as it reaches the edges, stretching into parts of Montana, Saskatchewan, and Manitoba.

This bowl is comprised of several layers of rock, with the source rock, or oil-producing layer (what we sometimes call "the kitchen") being on the bottom. The oil has been seeping up through the various layers for millions of years. In numerous spots throughout the basin, geological events—earthquakes, volcanos, etc.—have formed a mostly underground mountain range (Nesson Anticline) that extends north south for 75 miles right through the center of the play, which has helped to naturally fracture the rock. It has done what we oil companies spend millions of dollars to do, which is to release the oil locked inside.

This means that in those spots where geological events have fractured the rock, oil and natural gas are flowing more heavily and are more recoverable. And these places are where we want to be.

To track these areas, we create detailed maps of the play, updating them monthly. We're looking carefully for yield patterns in existing wells. Any area producing less than 400,000 barrels is, for us, not economical, and we'll pass on it unless it was drilled prior to the new fracking techniques then we will take a more in-depth look. Otherwise, it's simply not worth our while to buy it and drill it.

A lot of it has to do with the operators, too. We are non-operators in the Bakken; our focus is on acquiring properties in the development and growth stage and hiring reliable operators to handle the drilling and production. There are always trade-offs. The upside is that we are participating with some of the premier operators in the area; the downside to being a non-operator is that we do not control the timing of when the wells are drilled and completed.

And it's all about the frac. Some operators do more fracking stages than others. Some use porcelain instead of sand as a proppant. Some may still be using older technology. There are a lot of factors to figure in, and if we are going to do our jobs well, we have to stay on top of this constantly moving target. The average investor simply can't do the daily tracking work it takes to stay on top of a play, but whoever you invest with should certainly be doing this.

We constantly track production numbers and consider the dozens of factors going into those numbers to determine where the developing areas are, what's underperforming, and what's not yet on anyone's radar.

Non-Op Interest Focus

REI's business model of acquiring non-op interests in producing or drilling new wells can be explained by the perceived value of the following advantages:

1. **Increased Diversification** – This is achieved by spreading investor risk over a large number of producing wells. Instead of spending resources to determine new exploratory, or "edgy," areas in which to drill, which may entail substantial risk, REI focuses its efforts on locating producing wells in prominent areas in the core of the play.

2. **Increased Buying Power** – Receive the benefit of a large operator's buying power with decreased costs due to the sheer volume of drilling contracts. Also, operators who drill large numbers of wells have greater purchasing power for pipe and equipment to complete the well.

3. **Operator Experience** – REI essentially taps into the knowledge database of others for free. Relying on large operators with substantial experience in the region enables us to learn from others in areas such as geology, drilling and

extraction techniques, an understanding of infrastructure, transportation knowledge, and completion efficiency.

4. **Lower Capital Expenditure** – Acting as an operator requires a large amount of capital along with a high degree of risk. REI prefers non-op interests in certain plays because of proximity and available resources. REI acts as the technical and financial intermediary, sifting through and discarding projects that present a high degree of risk, or fall short of their unique economic model.

5. **Important Minimizer of Risk** – Mechanical failures occur. Though all risks cannot be avoided entirely, non-op risks are dramatically decreased by diversification. In the event of a mechanical failure, limited ownership allows for smaller losses.

No Lease, No Grease

As I've mentioned before, there really is very little land in the United States that is unexplored or not owned by someone. Whenever we identify one of those good neighborhoods and a piece of property where we want to drill, we begin the process to lease the property, or to buy a lease or production from an operator.

Why not buy the property outright? First, landowners in active areas like the Bakken, for example, or in areas where there are known oil mostly aren't interested in giving up their rights to the resources below ground. Especially in actively producing areas, as you'll see, there are too many benefits for landowners. Second, in most cases, we only want the resources in the ground, not the ground itself. When we're done producing the oil and gas, we want to close the well and restore the land to its previous state for the landowner to use for other purposes.

With a parcel of land in mind to lease, we'll research who owns the property, much like you would any other piece of real estate—go to the courthouse, run the title to determine who owns the minerals under the surface—and then contact the landowner and mineral owner(s) to negotiate the lease.

We can only do this once we know the neighborhood—how are prices in the area? What has the neighborhood's production been like? How long has it been producing? What are surrounding leases going for? In the Bakken, land used to be really inexpensive—before it became the hottest oil-producing region in the country. These days, one acre of land in the Bakken will lease for about $3,000-$5,000 per acre, and some really strong properties may reach $10,000 to $30,000 per acre.

Then we'll contact the owner, often known as the mineral owner, and express interest in the property. With horizontal drilling we'll typically negotiate a 3- to 7-year lease with a 75 to 81.25% net revenue interest, or NRI. Sometimes, we will purchase federal leases, or leases owned by the U.S. government, through an auction process. Usually, a federal lease is a 10-year lease with an 87.5% NRI.

Put simply, when the land is leased for drilling, the mineral owner agrees to let the oil company conduct drilling operations on the land, making the company responsible for 100% of the expenses, or working interest. But in addition to cash off the top for the lease of the property, the mineral owner also generally gets 20 to 25% NRI, meaning he receives 20 to 25% of the revenue on any resources sold, while we earn 75 to 80% of the revenue from the oil produced on that land.

Additionally, mineral owners will often tie us down to a drilling commitment. A mineral owner wants to begin making his money right away, and this can't happen until we drill. Often, a lease will state that we must drill within the first year of the lease, or even within the first six months; otherwise, we'll have to pay additional money on that lease. Usually, if, within that five-year period of the lease, we haven't done any drilling, the lease will automatically revert to the landowner, and while we're no longer under any obligation to drill, we'll lose that acreage, as well as the money we originally spent to lease the property.

To address this commitment, some oil companies may drill shallow wells using Spudder Rigs, which are light drilling rigs usually used at the beginning of drilling a well to bore the hole. This serves as a kind of placeholder, while they gather the resources needed to drill in earnest. Landowners have gotten savvier, though, and will often stipulate not only when companies must drill, but how deeply, too, and will only let lessees hold their leases in the deepest-producing zones. Such practices might also lose my company the right to some land in the Bakken.

If, however, we have drilled and production has begun, the lease usually can be held by production, meaning the lease continues as long as we're pulling oil out of the ground, or until we begin taking steps to extract ourselves from the lease.

Large landowners and the federal government (for government-owned leases) usually will also add a clause called the Pugh Clause. This is the language used in an oil-and-gas lease to spell out what will happen to the portion of the acreage you have leased that either does not contain a well or is not included within a producing petroleum pool or unit.

The typical Pugh Clause reads as follows:

> "If at the end of the primary term, a part but not all of the land covered by this lease, on a surface acreage basis, is not included within a unit or units in accordance with the other provisions hereof, this lease shall terminate as to such part, or parts, of the land lying outside such unit or units, unless this lease is perpetuated as to such land outside such unit or units by operations conducted thereon or by the production of oil, gas or other minerals, or by such operations and such production in accordance with the provisions hereof."

While the clause may seem complicated at first, its meaning is quite simple: At the end of the primary term, the lease will expire as to any part of the land that is not being used by the petroleum company. Without the Pugh Clause, if your lease covered 600 acres and the petroleum company only put 20 of those acres into a pooled unit for a producing well, the lease would remain in effect for the 580 acres that are unused as well as for the 20 acres being used. Even though there is no production (and thus no profit) being extracted from the 580 acres, they would remain tied up by the lease indefinitely. However, with the Pugh Clause, the 580 acres would be released from the lease at the end of the primary term. You would continue to receive royalties from the production occurring on the 20 acres, and the 580 acres would be available to lease to another company when one comes along.

Depending on the additional language in the lease, you must keep in mind that, in this example, the petroleum company may be able to hold on to the 580 acres through methods other than actual production of oil or gas. Some of the typical ways that land can be maintained without oil-and-gas production are seismic work, continued drilling operations, or additional rental payments. As with all provisions of the lease, the Pugh Clause and any additional language must be worded very carefully to provide maximum opportunity for the petroleum company while also providing maximum protection for the landowner.

To see how this looks in practice, let's look at an example:

> My oil company has identified a property in North Dakota that looks promising. The oil company works with a landman, or a professional who brokers such leases, and he does some research into the property to learn that Mr. Tom Gold owns it. He also researches the surrounding area and learn that Mr. Gold's neighbors are leasing their properties at $1,000 per acre (this is just an example—chances are, it would be much pricier than this in the Bakken of North Dakota!).

> The landman representing the oil company approaches Mr. Gold about his land, and he tells him that he is interested in leasing 1,280 acres of his property. They agree to a $1,000 per acre lease deal, with Mr. Gold receiving a $1,000 cash bonus per acre, or overriding royalty interest. So right off the top, Mr. Gold gets $1.28 million for the lease. He could get as much as 20% NRI, or what the industry may refer to as a royalty interest. My oil company will put up all the money to drill the well, so Mr. Gold doesn't have to pay a cent, plus he'll get 20%

of all the revenue generated from what comes out of the ground.

We negotiated a 10-year lease that stipulates we will drill the first well within the first year, or else my oil company will lose the lease and the bonus money it has paid. The company and its partners drill within two years, eventually developing 14 horizontal wells on the property—eight wells in the Bakken Formation and six in the Three Forks Formation. It costs the company $9.5 million (typical for the Bakken and Three Forks) per well, or $133 million total, and the operator, along with the pool of investors in the lease and drilling program, is entirely responsible for those expenses. It takes two years to completely drill, install equipment, and place the wells into production. We retrieve 100,000 barrels of oil at the end of the first year, per well, or 1.4 million barrels total. The oil sells for $100 per barrel, which brings the total revenue to $140 million in the first year.

That means that within three years, on just that 1,280 acres, my oil company and its investors have earned $112 million (80 percent of the total) in gross revenue before operational cost and taxes, and the landowner has earned $28 million.

Let's say that over the course of the lease, each well pulls 500,000 barrels of oil out of the ground. That means that for those 14 wells, the company has produced 7 million barrels of oil. At $100 per barrel, this parcel of land has generated $700 million in gross revenue; that's $560 million for those with controlling interest on the lease (my oil company and its investors), and more than $140 million for the landowner, not to mention the money for the lease itself, which is $1.38 million ($1,000 per acre for 1,280 acres).

So, as you can see, landowners stand to make incredible profits in the Bakken and other active shale plays, and so can an oil company and its investors.

The Lease and Direct Participation Fund

This is where you, the investor, come in.

In traditional real estate, you can borrow 70 to 80% of the value of the property (and I have heard of some cases when it's been 90 to 100%) from a bank, then finance the property. In the oil-and-gas business, it's quite different. A company can only borrow up to 40% of the proven reserves—if you are purchasing production. However, if a company is putting together a lease on a property that isn't currently producing, it can't borrow the money from traditional lenders. So one way a company can raise capital is through a direct participation program, in which it forms partnerships with individuals or small investor pools to fund the leases and production.

Between 2005 and 2011, the number of taxpayers who reported adjusted gross incomes of more than $1 million in North Dakota nearly tripled. About 90 percent of the drilling in the area occurs on private land. In 2011, The New York Times featured a story about a retired rancher who earned $80,000 per month for his small share of mineral rights.

Source: Blaire Briody, "11 Shocking Facts about the North Dakota Oil Boom," June 6, 2013, Fiscal Times; A. G Sulzberger, "A Great Divide Over Oil Riches," Dec. 27, 2011, The New York Times

A company can put its leases together in a couple of ways. It may start by identifying a certain property or area of interest, putting the land together first and then taking that property to the investors for a drilling. Or the company may start with a pool of investors who are ready to invest in oil and gas, then beginning to seek out land and approaching the landowner. In other words, sometimes the land comes first, and sometimes it's the investors.

Most often, though, the oil company will begin putting a land deal together first. At REI, we have acquired properties with existing production and upside developmental drilling opportunities, in order to minimize our risk. If we are leasing ahead of where future wells are to be drilled in the play, our leases are held by production, so we're not under the gun to drill a well; in most cases, we can borrow up to 40% of the money needed for this. We prefer this approach because of the high costs of buying the reserves in the ground, leasing the land, and drilling the wells. In areas such as the Bakken and other horizontal plays, we will participate as a non-operator, and we are sure to partner with the best operators in the business to handle the actual production. (I'll be discussing operators and non-operators in the next section.)

However, in the case of a lease—such as the example I discussed previously of Mr. Gold's land—an oil company raising money from investors would take the deal it had discovered to private investors in hopes of funding the leases and drilling the wells. One investment vehicle available to the company is to approach large and small investment firms, as well as, in a Regulation A offering (inspected and registered by the SEC), accredited and non-accredited investors. These investment firms would receive commissions to assist in raising the capital.

So in the above scenario with Mr. Gold's land, with that lease in hand, the company would approach accredited investors—those investors, alone or jointly with their spouses, whose net worth exceeds a million dollars, exclusive of their homes and automobiles—and non-accredited investors, whose net worth (alone or with spouse jointly) is less than $1 million, and whose net income has been $200,000 annually for the last two years ($300,000 with spouse). According to Regulation D, only 35 non-accredited investors are allowed to invest funds in a private placement.

If my company were putting the investors together through an S-1 offering, we would take 100% working interest, on which we own 80% NRI, to investors who want a piece of that. Because we're brokering the deal, like a realtor, we will take a 10% carrying interest on the lease. So we have a 10% working interest, as well as net revenue interest on that 10%. And for each investor's working interest, he or she gets NRI as well.

The timing on this disbursement of revenue generated from the sale of oil and gas is generally monthly, but it depends on the deal, when the oil is produced and sold, and other factors. Typically, the first revenue check arrives in 45 to 60 days from the date of first production.

Investors need to do their homework. When a new well is drilled, the amount of oil pulled out of the ground is at its highest rate right away, and it eventually levels off. This is especially true in the case of horizontal wells. This is because when you hydraulically fracture a well, you are using pressure, water, and sand to break up the rock and blow it up like a balloon. So, naturally, the oil, gas, and water flow very quickly in the early months of production. It's important to know at what stage of development the property is, in order to know how much production to expect and when.

Obviously, when you find reserves, land value will increase. This is how people who owned simple farmland have become millionaires. Between 1953 and 2008, the Bakken shale produced a total of 135 million barrels of oil. But in 2008, unconventional drilling took off in a major way, and today it produces over a million barrels of oil a day. In six months, it produces what the entire field produced in its first 55 years of life. As a result, land values have increased significantly. In 2010, the average price per acre for Bakken land was $300-$600, and by the middle of 2013, it was starting at $2,000 per acre and going as high as $30,000 an acre in some areas. Today, very little open acreage is left that does not have at least one well on it. So in order to drill a new well, one would have to purchase the existing producing well.

Because production can so drastically change the value of the property, any lease should include an exit strategy that spells out how long you'll be involved in the property, and how you'll divest yourself of it when you're done. Some sell it off, and some sell portions of working interest to other companies. REI likes to purchase working interest and develop wells, and we usually begin our exit strategy in three to five years. The goal is to sell when there's still "some meat on the bones"—when production is at its peak and/or oil prices are high, when we can sell for the best returns for ourselves and our investors. I'll spend more time discussing the end game in the next chapter.

Obviously, any type of investment has risk, no matter what it is. But if there was ever a perfect time to invest in production, it's now. Oil production, as I've explained, is cyclical. Currently, prices are relatively high and the market has been volatile the last few years, due to the COVID-19 pandemic and subsequent rebound; as well as the war in Ukraine; increasingly rigid environmental, social, and governance standards; and general inflation in all sectors of the economy. As *Kiplinger* pointed out in fall 2023, "With all due respect to Charles Dickens, the oil and gas sector is increasingly a tale of two markets. Depending on where you focus, it is the best of time, and it is the worst of times."

However, since the pandemic, as I mentioned earlier, oil-and-gas companies have held back from new drilling as a way to repair their balance sheets, meaning that they're now much less sensitive to volatility than they were a decade ago.

As *Kiplinger* also points out, the world's demand for energy is growing by the day, while the costs involved in drilling have climbed as a result of inflation. The rule of supply and demand is the first rule of economics: A commodity with limited supply and high demand means that its value increases. This is why the time is ripe for investing in oil and gas—particularly in light of its many tax benefits.

Plus, crude oil prices have risen over 90% in the last three years or so. As a result, stocks tied to crude oil prices are generating incredible cash flow. That's why now is when you should be looking to invest.

We're always going to need oil. There will be bumps in the road on any investment, but if you stay the course, you're likely to see the benefits.

REI ENERGY PROPERTY ACQUISITION MODEL

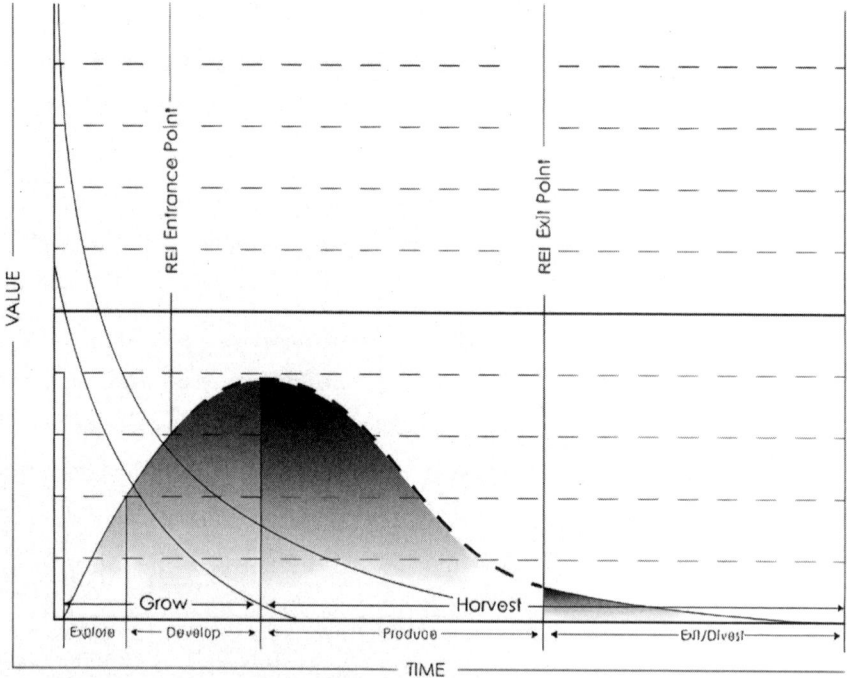

Lease Fund Proceeds Disbursement: A Case Study

Let's take a look at an example of how all this works.

ACME Oil & Gas Company has purchased interest in a 5,000 gross acre property in a new unconventional play in Oklahoma that already is home to 45 producing wells. ACME intends to triple the number of wells on the property upon purchase to improve upon the value of the land and the potential revenue for investors. ACME's business model is to participate in the drilling of 73 horizontal wells in three years, then sell the property after increasing the cash flow and value of the acreage, for a potential profit based on a 2.9-to-1 return.

Now, the company has put together a lease fund offering valued at $5 million, which it is calling "ACME Private Drilling Fund." This fund offering identifies a target number of 50 subscriptions—partnership interests of $100,000 each—being made to potential investors. The offering stipulates that potential partners must be high net worth individuals who satisfy the requirements of "Accredited Investors," ensuring financial suitability for the investment.

As part of the deal, the company reserves the right to expand the offering to $10 million, which opens up 50 additional units. The offering price per partnership unit subscription is $100,000; the minimum investment allowed for the offering is $25,000, which purchases interest in one quarter of a unit. The offering is being made available on February 15, and the offering period will terminate on July 31 of the same year.

Also stipulated in the offering are the risk factors and details about how cash distributions will be handled in terms of profits and losses. The offering states that if there are cash distributions to be made monthly (if the wells are producing and bringing in a profit), investor partners will receive 89% of proceeds, and ACME will receive the remaining 11%.

Mr. Clancy is an accredited investor who has been approached by a broker with ABC Financial about this potential investment opportunity. The prospectus on the property looks promising, its track record is favorable, and the existing 45 wells on the property are producing as many barrels as expected or more. Mr. Clancy knows about the many tax benefits of oil-and-gas investment.

He has also been told that ACME Oil & Gas strives to generate returns of between two or three to one on initial investments over the course of three to five years, and with the amount of producing wells expected to triple over the life of this fund, this seems feasible.

Mr. Clancy decides to invest $100,000 in a partnership in the ACME Private Drilling Fund.

The fund is viewed favorably by many investors, and the offering is expanded to $10 million, with 100 subscribers investing in the fund.

The upfront cost is 15% of this investment, to cover fees and commissions (typically, brokers take 6-10% on a deal). This means that of Mr. Clancy's $100,000 investment, 15% of it, or $15,000, is immediately taken by ACME to cover overhead costs, legal fees, administrative costs, and commissions.

The remaining $85,000 is what actually goes to the ACME Private Drilling Fund. However, because the initial investment was $100,000, this is still the payout amount, regardless of how much actually goes to the property; distribution percentages are still based upon this $100,000 investment.

Because this property was producing oil at the time it was funded, distribution payments are made immediately to investors each month. Of the total revenue generated, 89% of that goes to investors; it is divided amongst the 100 units, or investors, based on initial investment amounts, and those checks go out each month to investors.

Mr. Clancy's monthly disbursement checks are different each month. Early on, they begin as a small percentage of his investment, due to the fact the fund has invested capital in land that is not developed as the new wells are drilled, and typically, because the wells are horizontal wells, they will produce heavily in the early years and generate high cash flow as new wells are brought into production. Meanwhile, he also is receiving significant tax advantages (see Chapter 6) that make this investment advantageous for him.

Over the course of three years, Mr. Clancy has received these disbursement payments each month, plus reaped significant savings through tax write-offs. The fund sells its interest in the property. Its cash flow, which took just 36 months, or three years, to reach returns of three to one. This means that Mr. Clancy has not only recouped his original $100,000 investment, but he has made an additional $200,000, all in just three years.

The Costs of Exploring and Drilling

As an investor, you probably already understand the law of averages. In other words, you'll simply stand a better chance of seeing a return on your investment (and significant cash flow to boot) if your property contains multiple wells.

As I explained in Chapter 5, although conventional wells, in most cases, are less expensive to drill, a company might need to drill three or four conventional wells to find oil, and it's still not assured of making back its investment. Meanwhile, unconventional wells cost more, but typically they are more productive; if you're drilling in the right areas, you'll get oil every time. So with several wells on a property, you're spreading your risk because reserves can vary widely from one location to another, due to the composition of the rock and the potential of unforeseen mechanical issues and cost overruns while drilling.

One factor that will affect your investment, and which you should understand prior to entering into a lease fund, is whether the company you're investing with is an operator or a non-operator.

Operators physically oversee drilling. Similar to a contractor who is building a house—handling the plans and making sure they're followed, hiring skilled tradespeople, making decisions about materials or construction challenges, and occasionally doing some hands-on work—the operator handles production of oil and gas on a site. Operators are responsible for buying materials, contracting drilling rigs, and coordinating the hands-on efforts of pulling oil out of the ground.

For the last 25 years, REI Oil & Gas has conducted business as an operator. But in the Bakken, where we are involved in a large number of wells, we've chosen to have non-operator status.

In the Bakken, large public companies learned early on that unconventional drilling was expensive, both in materials and manpower, and the only way it was economical to do it was to buy and drill in bulk. With the drilling fever that hit the Bakken a decade ago, the equipment was tied up by these large companies early on. Rather than try to compete for that equipment, we determined that it made more sense to purchase working interest from existing deals and let existing operators handle production.

Through this experience, we've mostly bought interest in properties that were under operators, ranging from great to terrible. Some we want to continue doing business with, and some we never want to work with again. But before we enter into a play, we look closely at the operators involved, considering the costs of their operations and whether they typically are involved in good areas and productive wells.

If we buy interest in a well, it's just like buying interest in a business: We are responsible for a percentage of the costs of doing business, which means that we have to be smart about who we involve in the day-to-day work of that business.

As a non-operator, we don't have control over when the wells will be drilled. But if we're involved in deals with good quality companies, we don't have to worry about the logistics of putting wells together; we just get to ride along with the bigger operators, and we get the same amount of production.

Many wells in unconventional plays are owned by an operator and a host of non-operators. This means that when you invest, you want to know whether the oil-and-gas company is an operator or non-operator. If it's the former, is it a quality operator that is cognizant of keeping costs down and getting the best price for the product? If it's a "non-op," the company you are investing with should know and have researched what kind of operators it works with. What's their track record? This is the kind of homework a good investor should be doing. It's essential to be sure you're in with the right people—that they provide top-quality service and produce in a prudent manner.

Investment Risks

Of course, when you invest in any commodity, be it oil or gold or even agriculture, you are taking on a certain amount of risk. Your investment is subject to the whims of the marketplace and what the commodity is worth—its sale price—as well as the costs of extracting the commodity. In the case of oil and gas, these costs include buying materials and equipment and employing workers.

There's an old saying that goes, "Oil men go to Las Vegas to relax." That's a lighthearted way of saying that the oil-and-gas industry is prone to uncertainty with regard to oil reserves, exploration, crude prices, product prices, supply, and demand. Additionally, there are the costs of employing workers as well as the costs of materials and equipment needed for drilling, which may change due to other factors.

To mitigate these risks, which can be considerable, some companies will buy into turnkey projects. This means that the leases begin after the drilling phase has begun; the drilling costs have already been paid for, and there's no need to absorb them. Of course, you pay a premium for turnkey properties, but it minimizes a certain amount of risk. You just have to be careful with such properties because some companies will charge high premiums to cover their downside in case they run into problems; this could have a severe impact on the potential returns for investors—especially in shale plays.

In terms of product price, investors can, and often do, buy hedges, which are exchange-traded funds (ETFs), to protect themselves from the volatility of the market. Essentially, hedges are bets that investments will move in certain directions. Oil companies can do this in several ways:

- **A swap:** In this hedge, oil companies estimate a certain amount of production each month, and then there's a "swap" in terms of delivery of that production amount. In other words, if we estimate 1,000 barrels will be produced each month, and the property only produces 985 barrels, the leaseholder(s) must pay the bank the difference for the missing 15 barrels.
- **A ceiling or a floor:** These hedges are used to protect from price volatility. A ceiling, also known as a "call," is a cap on

how high prices can go before the investor is exposed to the risk of having prices rise too high for the marketplace. If I invest when prices are at $100 a barrel and they rise to $125 a barrel, I'm exposed to risk because demand will go down. Meanwhile, a "put," or a floor, is a minimum price per barrel that you want to ensure you don't go below. In other words, if oil is $100 a barrel when I enter into the lease, I might invest in a floor of $85 a barrel. If the price drops below $85 a barrel, I'm guarded against the loss I would incur from the drop in profits.

- **A costless collar:** Buying a collar hedge involves a ceiling and a floor. I'm protected from the upside and the downside. This sounds like a win-win, but remember that a ceiling caps prices, meaning that I could be missing out on increased profits if demand doesn't fall significantly.

The above risks are commercial. But also inherent to commodities such as oil and gas are global risks and element risks, both affecting upstream and downstream phases of the industry.

In Chapter 2, I explained why there are global risks involved in investing in oil and gas. Political, legal, commercial, and environmental factors are constantly at work to affect the costs of oil-and-gas production. These include changes in presidents or national leaders (here and abroad), international restrictions, the exchange rate and how currency in various markets is valued, and natural disasters.

Elemental risks involve the costs of recovery. These include construction, operation, financing, and revenue generation, as well as crude oil characteristics.

Your margin can be significantly affected by crude oil characteristics—whether it's sweet or sour. Not all oil is the same.

The American Petroleum Institute (API) measures viscosity of oil with a measure called "gravity," which compares crude weight to that of water. The heavier the crude, and the higher its sulfur content ("sour"), the harder and more expensive it is to refine and make useful. Sweet crude is lighter and low in sulfur content, making it easier to refine. The price of oil you will hear quoted usually refers to sweet crude.

So if the crude coming out of the ground is more expensive to refine, it could mean less money for the company and its investors.

The location of production and transportation of the oil and gas are also important recovery factors. The closer the production is to the pipeline, the cheaper it is to ship. On the other hand, production that's farther away may have to be trucked or placed on a rail car, which adds to costs.

As you can see there are numerous factors that can affect oil-and-gas investment, which all increase the risks investors take. This is why REI provides a tremendous amount of information to any potential investor. In order to secure this kind of investment, we must offer full public disclosure.

Generating Cash Flow

Bottom line, what we're talking about in this chapter is cash flow—all the factors to be considered when you're making an investment in oil and gas, with the goal of generating the most cash flow.

That's why we'll only pursue properties that offer a certain amount of proven, recoverable reserves. This means that the reserves have been established by geologists and engineers and have up to 90% likelihood of being viable for extraction, which is the highest degree of confidence available. This is in contrast to probable reserves, which are likely to be recoverable, but the confidence is much lower; probable reserves have at least a 50% likelihood of providing recoverable assets.

It's not enough for a well to be producing; it has to produce enough to make the money we put into the ground worth it. Unconventional plays, which comprise the bulk of our work, come in at high production rates and then drop off relatively quickly to level out. That means that we want to be on the front end of the process, so we get our upfront cash flowing back. We're looking for wells that will produce between 60,000 and 100,000 barrels in the first year, which would be a "flush production," or enough to get most of our money back.

So as an investor, how do you calculate production into your cash flow equation? You have to look at how much money is flowing away from you, your costs, in order to know how much is flowing toward you. Each deal and each investment is different, so you'll need to carefully consider all factors involved, and be sure you have reliable information to determine whether the cash flow makes the investment worth it for you.

If you're looking for significant, steady cash flow, and to make back that initial investment, you'll have to understand what kind of money you're talking about, and whether that amount makes it worth the risk for you. Be sure that you have an investment professional whom you trust to run these numbers for you to help you make such decisions.

It's also important to point out that there are a lot of scams out there for oil-and-gas investment. Particularly when oil prices are high, as they are now, many investors try to capitalize on the possibility of making money, and con artists often try to take advantage of this.

For example, a scam artist might approach you with an unsolicited investment opportunity, to get in on the "ground floor" and help fund an oil well drilling project that promises to make a lot of money. The person who contacts you may try to flatter you or convince you that you are one of an elite few to be chosen for this opportunity. Aside from flattery, another tactic used by scammers is a sense of urgency; we're often unable to think through an idea when we're told we have to act quickly.

Another hallmark of a scam is the promise—a guarantee, in fact— that you'll earn incredible returns. If it sounds too good to be true, it probably is. High-pressure salespeople, who may have little or no expertise in oil and gas and are simply good at selling, may try to apply pressure on you to act fast before the opportunity disappears.

Some scammers may even provide you with slick-looking printed materials that lend an air of legitimacy to the deal. But a savvy, knowledgeable investor would never make a large investment based on a cold call and a brochure.

In some scams, the business offering the investment is located in one state, the drilling operation in another, and the shares are sold in all but those two states. The reason for this is simply that an investor can't drop by to look in on the investment if its operations are in other states.

It's critical that you invest with a company you can trust and that there are no barriers to the information you need to make an informed investment decision. Be sure you conduct thorough research on any company making investment offerings. Understand its business, the regulations it's subject to, its financial statements, and more.

Analyzing Potential Investments: A Case Study

Ms. Randall is an Accredited Investor who has talked to her financial advisor about oil-and-gas investments. She has learned of the many significant advantages of this particular type of investment, including:

- Intangible Drilling Costs (IDCs)
- Tangible Drilling Costs
- Lease Costs
- Intangible Completion Cost
- Depreciation
- Depletion Allowance
- Long-term cash flow

She is interested in making such an investment in the development of wells as a way to gain all the tax advantages available to her; she knows that the cash flow from such an investment can last a decade or more, and she understands that she can write off nearly her entire investment. She has $100,000 that she is ready to invest, and, after consulting with her tax advisor, she discovers this formula to calculate the potential return on investment:

$100,000 investment x 70% (IDC deduction) x tax bracket percentage = Real dollar savings

Here's how this applies to Ms. Randall:

$100,000 (investment size) x 70% IDC deduction
x 37.5% (tax bracket) =

$26,250 ESTIMATED TAX SAVINGS

THAT'S RIGHT: *THIS REPRESENTS A 26% RETURN ON INVESTMENT FROM THE TAX BENEFIT ALONE!*

Knowing this potential benefit, Ms. Randall is eager to invest in the right opportunity. After a few months, an oil company presents her with an offering for her consideration. She must analyze the investment to determine whether it is right for her. She has been given a prospectus to review. She knows she must analyze several factors before making her decision: geology of the area, the structure of the deal, and the company making the offering.

She can see that the property is located in the Bakken formation of North Dakota, a hard-rock area that has a strong reputation for consistent oil-and-gas production over the last decade and a half. While certain areas of the country present significant risk in terms of stability (Southern Louisiana, for instance) or in being "wildcat" areas (regions of exploration), this area of the country has been consistently proven to produce oil, and much of it, geologists have said, is still yet to be tapped.

The company making the offer has indicated that it strives for a payout period of three to five years. However, up until recently, the price of oil was more than $100 a barrel. At the time she receives this offering, however, the price is closer to $750 per barrel.

Ms. Randall is a savvy investor, and she has asked how much it costs to produce a barrel of oil and finds out that it's about $35 a barrel.

She has read the company's Private Placement Memorandum (PPM) and knows that the well must produce 200,000 barrels of oil for her to break even. Through her research, she understands that horizontal wells in an actively producing area can generate as much as 50,000 barrels a month for the first several months and up to 50% of their reserves in the first three to five years.

If a well is estimated to produce 800,000 to 1,000,000 barrels of oil over its lifetime, it can yield 300,000 to 500,000 barrels in the first several years. This means she can potentially double her money.

When Ms. Randall examines the area in which the company is drilling, and she notices that several of the surrounding wells produced oil and gas, but they appear to only be marginal producers. Upon asking the company's representative why it is drilling in a poorly producing area, she learns that the offsetting wells were drilled about 10 years ago in the early stages of the field development, using old technology. There have been numerous advances not only in the drilling of wells but also in fracking, which allows companies to go back into areas that were once thought to be poor and, through the new technology, allow them to recover as much of the reserves as possible.

Ms. Randall can see from some of the newer wells in the area that they are performing much better than the older producers, and through her analysis, she begins to believe that the cash flow and tax benefits, based on a $100,000 investment in this offering, make such an investment worthwhile. And, if prices hold where they are or increase, she can potentially do much better than the company is projecting. She decides to invest her money in this fund.

CHAPTER 8

THE END GAME

"My formula for success? Rise early.
Work late. Strike oil."

— AMERICAN INDUSTRIALIST AND OIL MAGNATE J. PAUL GETTY

So how do you know when you've found the right energy partner? How do you evaluate the company behind the deal to determine whether you've found a legitimate deal that will benefit all involved, or just a scam to take your money? After all, you can't truly calculate what kind of money you'll make unless you can feel as confident as possible that the assumptions made by the oil company will be correct—the projections about production, the recoverable reserves on the property, and the value of the net assets after costs are taken out.

The first thing to remember is that if the company hesitates to provide information about its company's history, development team, company officials, or production numbers; if you have difficulty obtaining it; if nothing is provided to you in writing; or if the numbers simply don't make sense or are presented in an overly complicated way, you should *run*, not walk, away from the deal.

You must be allowed, and encouraged, to do your due diligence. At REI, we make it our business to do our own due diligence on any potential project. We would expect no less of our investors. We only want investors who are fully informed and knowledgeable about where their money goes. To ensure this, we provide a tremendous amount of information to potential investors through a prospectus, and anyone you consider investing with should do no less.

Every investor must analyze the investment opportunity carefully—the deal, the oil company involved, and the broker-dealer presenting the opportunity to you. You should fully understand with whom you're investing, why you're investing in oil and gas, and why you're investing in this project. If you have trouble finding this information, that's a big red flag.

Scams, unfortunately, run rampant in the oil-and-gas business. To safeguard yourself, make sure you get everything in writing, and be aware of any opportunity that is promised to make you rich, or which is presented as "too good to pass up." Also, beware of high-pressure tactics that force you to rush a decision, or that indicate this as a "limited" or "once-in-a-lifetime" opportunity. And last, no matter how good the company is and how much due diligence a company does, there still is a chance you can lose all or part of your investment. If you cannot bear the economic loss, you should not invest.

Brokers are required to produce a private placement memorandum, or PPM, that details every project and deal. While it's their job to understand this information and perform due diligence, it's your job as an investor to be smart with your money. Understand how to read a PPM and what the numbers provided mean in order to determine how much oil or gas must be produced in order to make money on the project.

Choosing the Right Energy Company as Your Partner

Making the right choice about with whom to invest can mean the difference between significant cash flow and going bust.

In the Bakken alone, there are about 50 large to medium-sized companies working as operators—this doesn't even factor in the number of companies, like REI, that are non-operators with investment holdings in this area.

With that many companies, how do you choose the right one to partner with as an investor? You're looking for good cash flow and manageable, acceptable risk.

Beware of any company saying things like, "We're in the Bakken/Permian/one of the unconventional plays that are the buzzwords in our industry today; let's drill a hole! There are no dry holes—you'll make money!" This is not grounds for you to invest. As you now know, while these areas may have abundant reserves, there is *never* a guarantee that an oil well will produce oil. And if you do find oil, it's still possible that the ends won't justify the means.

Here are a few of the key criteria you should evaluate when looking at any oil-and-gas company:

Company History and Track Record: What do you know about the company? How long has it been in business? How large and how knowledgeable is its team? Who is at the helm, and how experienced are its leaders? Where are its headquarters? Is there consistency in its decision-making process?

Look at the company's drilling efforts. Where does it focus these efforts, and how many wells does it have stakes in? Is it involved in the most prolific areas for oil-and-gas production?

Is it an operator or a non-operator? If the latter, who are its partners, and what is their track record? A company's track record can be deceptive in today's oil-and-gas industry because every well in one of the hot shale plays around the country will be completed by a producer. So nearly every company can tout almost a 100% successful track record. But where the rubber meets the road is in understanding where their acreage is located in the play. Remember, it's a lot like investing in real estate: location, location, location.

What about transportation? This can have a huge impact on the revenue earned; the operator must have a method in place for getting the crude from the well to distributors and, ultimately, the consumer. Transportation can determine the speed of payout, price per barrel, and more. Will it be distributed by pipeline, trucks in tanks, or rail? According to engineering.com The biggest advantage of pipelines for the transportation of fluids, oil, or otherwise, is the sheer efficiency. Think of a pipeline as the carpool lane on a freeway or a railway system without stops. There are no physical barriers to impede its transit. Additionally, because oil fields can be found in remote locations, railways will likely be farther from the source. In turn, pipelines are often the most cost-effective way to move millions of gallons of oil across thousands of miles of land and water. As a result of this dedicated expediency, it costs much less to transport oil by pipeline—about $5 per barrel, compared to $10 to $15 for rail transport.

Rail has been a major innovation in the Bakken, as it enables operators to ship from the Bakken to Cushing, Oklahoma, to a terminal that enables the oil to go to the Gulf Coast, the East Coast, or the West Coast. Where operators can get the maximum price per barrel. While pipelines have fixed locations, rail transport can be more flexible for transporting oil and can deliver oil to and from a wider variety of locations on short notice, depending on market demand. Unlike pipelines, rail shipment does not necessarily require long-term contracts, nor is there a regulated rate of return. A pipeline, for instance, can't raise its rates when service demand is high. "A pipeline can get congested during peak times so you can't get any more oil through it, but the rate still can't increase above that maximum regulated rate," University of Chicago public policy professor Ryan Kellogg told *Energy News*. "So, the fact that pipelines can become congested creates the opening for rail to come in and help move the extra oil."

As part of our due diligence, REI looks at all operators with whom it works and explores how they distribute and how much each gets for its crude so that we can work to secure the best prices possible.

Here's something else to consider: A lot of public companies like to brag about the booked reserves they've found in certain areas. And that may sound exciting, but all that really matters is what they have to show for it. Reserves aren't the same thing as production. Are those reserves proven or probable? And are they recoverable? Will it cost $100 million to pull $10 million worth of oil out of the ground?

This is why it's critical that you look at a company's past performance in the targeted development area. This says a lot about the production team as well as the area itself. County offices have information about wells being drilled within those counties, and this can turn up area production numbers; it can often be interesting to compare promised reserves with actual findings.

In general, you want a company that partners with major oil-and-gas companies and leading independents, and which consistently and economically produces oil and manages risk to provide good returns to investors. What is the company's exit strategy?

Technology and Innovation: In general, the newer the technology, the more effective the production. You want a company that has successfully employed the very latest in drilling and discovery technology, and which is well versed in the variety of technology in the industry and what's being used in the target area. And because technology is only as good as the people using it, you want to find a company with a seasoned team of land, geological, engineering, marketing, and finance professionals.

Investment Offerings: You want to find a company that offers breadth and depth in its investment offerings. It should offer programs covering the spectrum of risk tolerance and product diversification— from conservative to aggressive and everything in between. This approach helps spread risk and offers balance to your portfolio.

Communication: If you're going to put your money into the hands of an oil-and-gas company, it should be someone with unquestionable integrity. You want honest, thorough, open dealings and complete transparency. Be sure you understand the long-term strategy of any company you are considering as an energy partner. How you will communicate with the sponsor, and with whom? What level of liability insurance does it carry? What is the level and strength of its communication with partners? What type of participation does it offer to partners?

REI makes communication with investors a top priority. We frequently provide up-to-date information to investors about our current partnerships, and we employ a knowledgeable group of CPAs who can answer investors' questions about tax benefits, but each investor should consultant his or her on CPA. Plus, we offer a fully integrated website for downloading cash flow statements and tax information. All the information investors need is available quickly and easily, so investors are never in the dark about where their money is going.

Selecting the Right Broker/Dealer

In many cases, the person asking you to invest will be a third-party broker/dealer, who may also be referred to as a sales representative, stockbroker, account executive, or registered representative. This person acts on behalf of the sponsor (the oil-and-gas company).

Due diligence is essential when selecting a broker/dealer, too, because there are plenty of brokers out there looking to take advantage of people by pretending that they are more knowledgeable and trustworthy than they actually are.

The first thing you should check is whether the broker has the proper credentials and training to be handling your money. Is this person a member of the Financial Industry Regulatory Authority (FINRA)? Does this broker have an education in securities and investments, and is he/she experienced in handling large investments in oil and gas? While such credentials can never be considered guarantee of quality or of earning high returns, it does ensure a minimum level of knowledge that should help you in making an informed decision about investing.

Here are some things you should consider when selecting a broker/dealer:

Experience: Find out whether the broker/dealer you're considering is competent in the type of investment in which you want to participate. What is his/her education, experience, and credentials, and does he/she have proven success in this area?

References: In addition to seeking SEC and FINRA affiliation, it's a good idea to seek out someone who is registered with the appropriate self-regulatory agency (SRO) and be sure that he or she does not have a history of customer complaints.

Find out everything you can about the project: What types of reserves are being estimated, and how recoverable are they? Were they evaluated by a geologist and reservoir engineer.? What's the history of the well? Has anyone drilled there before? If so, how much was recovered?

Is a third-party engineering report available? The reservoir qualities of a play are studied by petroleum engineers, who are practical in their evaluations of such factors as downhole pressure, porosity, permeability, and many other things that help them to determine how much an oil-and-gas well could ultimately produce and what recoverable reserves will be found there. These engineers are often hired by banks to examine these factors and produce reports, and these reports become part of the due diligence reports, so be sure you review them carefully as they'll reveal valuable information to you.

There are a lot of companies who are not registered with FRINA trying to broker non-registered deals. While there is no 100% guarantee on any investment, working with a qualified and credentialed broker mitigates your risks and assures that you are dealing with someone who has been educated in offering securities. And it provides you with certain legal protections.

The Final—but Most Important—Piece of the Puzzle

Homework, or due diligence, is crucial in matters of investment. So is an understanding of risk and of the ins and outs of the oil-and-gas industry.

But never underestimate the importance of instinct. When it comes to any major investment, what is your gut telling you about the project, the oil company, and the broker/dealer?

Do you believe instincts can't be trusted? Research says it can. There was a psychological study done in 2012 in which participants were asked to choose between two options in a series of number pairs. Using instinct alone, and no instruction or materials to study, the participants made the right choice about 90 percent of the time, thanks to their instincts.

Trust your gut to tell you whether the project and the people involved inspire confidence in you and whether you feel comfortable and not pressured by anyone. Be sure that they pay attention to your needs and that everything is explained to you fully until you completely understand what's involved.

Common Red Flags

When presented with any investment opportunity, be mindful of these warning signs that can indicate a scam or less-than-quality offering:

- Sales pitches referring to recent news events like high prices for oil or gas.
- "Can't-miss" wells and "guaranteed" returns, including claims that major oil and gas companies are drilling nearby.
- Promises of abnormally high rates of return.
- Unsolicited materials being sent to you.
- Sales tactics that pressure you to decide in a hurry, like "limited" or "once-in-a-lifetime" opportunities.
- Sales pitches touting new technology, especially if it relates to getting higher production out of low-producing wells (sometimes called "stripper" wells).
- Salespeople who claim to also be investors.
- Requests to sign documents acknowledging that the securities laws do not apply to the investment.

Exit Strategy

As I mentioned earlier, there must be an exit strategy in place from the get-go, and this is one of the things you should consider as part of your due diligence process.

Because REI is a non-operator in the Bakken and other shale plays, we can't control the drilling schedule or process, but we also have enough experience in the Williston Basin, and have done enough extensive research, to know the best areas and who the best operators are, and we've invested in commissioning in-depth studies on the area to determine the most promising prospects. We have a technical team whose goal is to bring us 20 or 50 potential prospects per year; those teams are comprised of Land professionals, reservoir engineers, production engineers, completion engineers, and various other experts, all weighing in with their expertise about the properties. We move quickly to evaluate each of the opportunities and set out to participate in roughly 10 to 20% of them.

Once we decide to participate and have our partners lined up, it can take up to three to five months to get the permits and land required to drill a well; then, once drilling in the well begins, it will take an estimated four to six weeks to finish drilling it. Completing a well can take anywhere from one to three months, and as part of this process, the ground is stimulated to start the flow of oil or gas. This can be the beginning of a well that produces for 20 to 30 years.

Obviously, as soon as a well comes online, it becomes a depleting asset. You could hold it until it's depleted or divest of it once you have received the majority of the reserves. If you imagine that bell curve of production that I've referred to earlier, we like to exit at the top, at the height of production when plenty of reserves still remain. Often, we'll leave space on a drilling unit for another company to drill new wells later on. This helps to ensure that we have something worth selling, and that we can get top dollar when we exit. Of course, this is often based on the price of oil, which we've seen is something that's nearly impossible to predict. We might hold on to a property a bit longer if it means we can wait for a more favorable price and generate good returns for ourselves and our partners.

This exit strategy also relieves us of the obligation to plug and abandon the wells and remediate the surface which can be a costly process.

When a well is depleted, the equipment is removed from the hole, and the empty well is filled with cement. The pipes are cut off at about three to six feet below ground level so that no empty pipes are left sticking out of the ground. All the surface equipment is removed, and the pads are filled in with dirt or replanted. At this point, the landowner can use the land again, and hopefully no signs remain that a well was there.

Although unconventional wells make a much smaller environmental footprint than conventional wells, remediation of vacated wells is resource-intensive. If we can help it, we avoid being the last one holding the bag. Some other companies specialize in different stages of the well's lifecycle. They may be looking for quick cash flow with very little development involved and might be interested in buying from us. Other companies own stripper equipment; they buy cheap, defunct wells and specialize in working them over to tap production that is no longer easily retrieved through traditional methods. And some others like to get in at the end and explore additional layers of rock that haven't been drilled.

Whatever your specialty, knowing when to exit is just as important as knowing when to buy. This means having bright, knowledgeable people constantly monitoring your portfolio and determining that right moment to exit. This comes with finding the right energy partner.

CHAPTER 9

THE GAME CHANGER

"Energy forecasting is easy. It's getting it right that's difficult."

— GRAHAM STEIN, COFOUNDER
OF GREEN ELECTRICITY MARKETPLACE

The oil-and-gas industry often gets a bad rap. The sins of the few often outweigh the virtues of the many, at least where public perception is involved. However, I believe what we do is vitally important, not only for the nation's energy needs, but also for its economy and its people. Constant innovation ensures that we do this work in the best possible way. As we've seen, people in this country rely on us for oil that fuels their cars as well as nearly every other object they use in their daily lives, not to mention for the thousands of jobs the industry provides and the boost to our economy it offers. For investors who stand to benefit, or to lose, from oil-and-gas production, the way we do business is even more important.

One of the primary reasons I've worked in this industry for over 35 years, and why I still love it, is because it's constantly changing. The way we do business now is a far cry from how we did things back in my early days, but the changing nature of the industry is exciting. Every day, new technological breakthroughs change how we pull oil and gas out of the ground—and how much we're able to retrieve. Even as I write this, projections about future production are being adjusted as processes improve daily.

The game-changing technology of fracking and horizontal drilling aren't at all static; their effectiveness is constantly improving. Greater drilling efficiency, more cost-effective and environmentally friendly methods, and new well productivity have been the main drivers of this production growth, not simply a greater number of drilling rigs.

In April 2013, the U.S. Energy Administration projected that we would be a net overall exporter of natural gas by 2020. But by the beginning of 2018, we had already achieved this feat. In fact, that same year, the United States exported 3.6 trillion cubic feet of natural gas, which was the highest on record at that time. Since that, our exports have continued to grow. In 2023, the EIA predicted that total natural gas output would increase by nearly 0.3 billion cubic feet per day in the major shale basins, which would be a record output.

In most other countries, shale development simply doesn't have the potential that it has here in the United States, due to a number of factors:

1. The quality of the resources and where they're located in the United States makes them easier and more worthwhile to harness.

2. Major private ownership of subsurface mineral rights, often by surface owners, provides a strong incentive for development.

3. There is greater availability of independent operators and supporting contractors with needed expertise, as well as access to advanced technology, here in the United States.

4. We already have an existing gathering and pipeline infrastructure in place.

5. This country has greater public acceptance of hydraulic fracturing as well as related activities such as transportation of materials.

6. The United States has the availability and methods for disposal of water/wastewater.

Oil and gas may be finite resources, but for the next several decades, they will not only dominate the energy landscape, but because of new technology in drilling and hydraulic fracking, they will be coming out of the ground in greater quantities than we've ever seen before. There's never been a more exciting time to work in, or invest in, oil and natural gas development.

The Future of Oil

Predicting the future of oil and gas is anything but an exact science. You've seen by now how many factors can influence the industry, from technology to political strife or even natural disasters. But each year, the EIA comes about as close as we can come to projecting what the near future looks like for the production of oil and gas in the United States.

In its *Annual Energy Outlook 2023* (*AEO23*), released in March of that year, the EIA indicated that the U.S. will remain a net exporter of petroleum products and natural gas through 2050. Oil production in the United States has grown to about 13.2 million barrels a day as of October 2023, according to this report, which is up almost 900,000 barrels a day from the same month in 2022.

While domestic consumption of petroleum is expected to remain flat, the U.S. is still the largest consumer of oil, and international demand will continue growing rapidly. Although experts say that the U.S. has already reached peak gasoline consumption, it's not anywhere near falling off a cliff. "It will take decades for gas-powered vehicles to drive off into the sunset," said scientist Rob Jackson of Stanford University in a 2023 article in *Fortune*.

Meanwhile, motor gasoline is still the dominant light-duty vehicle fuel. And despite the growth in the electric vehicle market, it's still expected that in 2030, gas-powered vehicles will still comprise seven out of 10 vehicles, or 69%, of vehicles on the road. Cars and small trucks will use less energy, as a result of stricter auto efficiency standards as well as lifestyle changes. The pandemic shifted the way people work, with 12.7% of full-time employees working from home and 28.2% working a hybrid schedule of both at-home and in-office days. This shift has had a marked effect on gasoline consumption, with fewer people commuting to work on a daily basis. It's expected that nearly one-third of Americans will work remotely by 2025. Also interesting is the fact that, as Brookings reports, young people increasingly prefer not to drive, with fewer of them every year getting their drivers' licenses. All of this adds up to less demand, in the face of greater supply.

But while that may be the case here in the U.S., globally, the demand for gasoline will keep growing. Statista reported in 2023 that demand is forecast to climb to 27.6 million barrels per day by the year 2045.

We're exporting record amounts of refined products and are poised to export large amounts of crude—particularly light, sweet crude, the best oil on earth, which just happens to be found in huge quantities in America's shale plays. Thanks to an abundance of oil refineries that process far more than we need for domestic consumption, it will be of economic benefit to us to export the excess.

Light crude oil fetches higher prices in commodity markets than heavy crude, and makes less of a negative impact on the environment. It's also easier and less costly to process into gasoline.

None of this means that our coffers will be too full. Though our import quantities will be nothing like the amounts we've imported in the past, we'll import large quantities of oil—heavier crudes—for the foreseeable future, because it's less costly to retrieve and is cheap to process, thanks again to all those oil refineries we've already invested in. The EIA has been urging oil "swaps," trading our light oil for the heavy stuff, which we can find cheaply from Canada and Mexico (a barrel of West Canada Select heavy crude is $21 cheaper than a barrel of West Texas Intermediate light crude) and process for almost nothing. Such swaps are expected to continue.

Also, even though non-OPEC production has skyrocketed and we're importing decreasing amounts from OPEC each year, production in key Middle Eastern OPEC countries (Saudi Arabia, Kuwait, and the United Arab Emirates) has not slowed at all. According to the

International Energy Agency's chief economist, Fatih Birol, "Despite the shale revolution, the Middle East is and will remain the heart of the global oil industry for some time to come."

However, even though China's standing as the world's largest oil importer means that the Middle East—where per-barrel production is less expensive than the United States—will dominate the growing Asian market, America's prominence as a major international exporter, particularly to other areas in Asia, will only grow.

Natural Gas Poised OR on Track to Explode the Energy Market

The figures on oil production are nothing compared to what's happening with this country's natural gas production. The EIA predicts that U.S. domestic natural gas production will increase by 15% between 2022 and 2050. We hit a record high amount of natural gas exports in 2022, making it six years in a row that the United States has been a net exporter—nine years earlier than projected a decade ago.

Because it is a much cleaner, more efficient source of energy than coal, natural gas surpassed coal as a source of electricity, accounting for 33% of our total energy sources, second only to petroleum, at 36% (coal accounts for 10%, and renewable energy for 13%).

Though the energy picture is certainly changing as new technologies and power sources are tapped, natural gas will remain a dominant player for the foreseeable future. "Despite no significant change in domestic petroleum and other liquids consumption through 2040 across most *AEO2023* cases, we expect U.S. production to remain historically high as exports of finished products grow in response to growing international demand," says the report. The largest increases will be in the industrial sector (as much as 32%) and transportation (up to 8%).

The report also indicates that, relative to 2022, natural gas generating capacity ranges from an increase of between 20% to 87% through 2050. Furthermore, natural gas-fired heating equipment will continue to account for the largest share of energy consumption for heating in the United States.

The EIA projects lower prices for natural gas than in recent years, which makes it even more attractive as an energy source.

In addition to electric power, industrial and transportation use will drive growth in output of natural gas, says the EIA. The industrial sector is expected to become the largest consumer of natural gas, with an increase in usage for heat and power as well as feedstock for chemical industries. Companies using freight trucks are increasingly powering their fleets with natural gas, which will be the primary reason for growth in transportation usage.

Meanwhile, our rate of natural gas imports, nearly all of it from Canada, peaked in 2007 and has been declining ever since, though imports are still used to supplement our supply during the winter heating season, which is when demand peaks.

In fact, there's so much natural gas coming out of the ground here that it can't all be used. Most of this nonmarketed gas is flared—a combustion process that uses fire to dispose of gas—into the atmosphere, releasing CO_2, a less powerful greenhouse gas than methane, as a by-product.

However, new processing plants and pipelines enable producers to bring more gas to market or to convert it to compressed natural gas, allowing it to be stored for future use by vehicles and equipment. And increased consumption should lower the amounts of nonmarketed natural gas as well.

No Crystal Ball

Despite the thousands of predictions that appear in the media every day that offer predictions on the future of the oil-and-gas industry, there's certainly no crystal ball that will allow us to know exactly what will happen in the future. But I can tell you that things look very promising.

It seems this is always how it's been in the industry. All the way back in 1859, Bob Drake, Sr., founder of the Hydracrete Pumping Company, put a call out for men to work as oil drillers. In the HPC newsletter, Drake reported that the response he received was something to the effect of, "Drill for oil? You mean drill into the ground to try and find oil? You're crazy."

Surely there were plenty of people who kicked themselves later when HPC and other companies not only drilled but struck it rich. It's hard to believe that such doubts persisted, but then again, even just 20 years ago, many experts believed we were on the verge of running out of oil. In 1972, the *Bulletin of Atomic Scientists* predicted that the U.S. supply of oil would be depleted within 20 years. And in 2002, the *Index Journal* in Greenwood, South Carolina reported that the world's supply of oil would peak in 2010.

No one could have predicted that George Mitchell, out of sheer stubborn persistence, would show us how hydraulic fracturing and horizontal drilling could be done in a cost-effective way, or how much of a boon this would be to creating an abundant supply.

Who's to say that it will stop here? Who knows what next innovation we're on the verge of now?

Natural gas surpassed coal as a dominant source of energy, says the EIA, and its *Annual Energy Outlook 2023* shows gas usage continuing to climb through 2050 and beyond, with no end in sight. And our number of exports of this valuable resource keep growing every day. Plus, as I've shared with you in this book, continuing heavy reliance on oil and gas worldwide, along with skyrocketing rates of consumption for some of the most heavily populated nations in the world have combined to place the American oil-and-gas industry firmly at the top of the heap, with investors squarely sitting beside them.

Continued investment in new technology is what has gotten us here. History has shown that investment in oil and gas has pushed the industry forward in new ways, and there's no reason to believe that won't continue. Perhaps the next game-changing technology is right around the corner, and your investment could be a part of making it happen.

MEET MIKE MAUCELI

Michael Mauceli is the founder and chief executive officer of REI Energy, LLC, and its affiliates since 1987. In addition to being an author, Mike hosts a weekly podcast for "The Rich Dad Network" on YouTube to educate listeners on domestic and international energy events.

Under Mike's leadership, REI Energy has experienced remarkable growth. What started as a three-full-time employee start-up company in 1987 has now evolved into a significant player in the oil and gas industry. Today, we are a leading oil and gas production, producing property, and mineral acquisition company, serving the investment needs of oil and gas investors across the United States.

Mike's vision and business acumen have made him one of the most influential oil and gas investor community CEOs. With his determination, REI Energy will continue to offer energy investment opportunities that have the potential to generate successful results.

In the 37-plus years since, Mike's companies have participated in over 1000 wells in the Bakken shale in North Dakota and Cherokee Shale in Oklahoma and operated and acquired interests internationally, offshore, and in over 2000 wells in 11 states domestically. Under his direction, they operated a 2-million-acre concession in the Northwest basin of Argentina, drilling the 4th deepest well in the country (and the deepest well drilled by a Heli Rig), making it one of Argentina's most profound oil and gas discoveries. In Canada, the company operated a horizontal well in the Alberta foothills, believed to be the world's deepest at that time.

These notable projects and successes are a testament to our expertise and commitment to excellence.